Operator Theory: Advances and Applications
Volume 230

Founded in 1979 by Israel Gohberg

Editors:
Joseph A. Ball (Blacksburg, VA, USA)
Harry Dym (Rehovot, Israel)
Marinus A. Kaashoek (Amsterdam, The Netherlands)
Heinz Langer (Vienna, Austria)
Christiane Tretter (Bern, Switzerland)

Associate Editors:
Vadim Adamyan (Odessa, Ukraine)
Albrecht Böttcher (Chemnitz, Germany)
B. Malcolm Brown (Cardiff, UK)
Raul Curto (Iowa, IA, USA)
Fritz Gesztesy (Columbia, MO, USA)
Pavel Kurasov (Lund, Sweden)
Leonid E. Lerer (Haifa, Israel)
Vern Paulsen (Houston, TX, USA)
Mihai Putinar (Santa Barbara, CA, USA)
Leiba Rodman (Williamsburg, VA, USA)
Ilya M. Spitkovsky (Williamsburg, VA, USA)

Honorary and Advisory Editorial Board:
Lewis A. Coburn (Buffalo, NY, USA)
Ciprian Foias (College Station, TX, USA)
J.William Helton (San Diego, CA, USA)
Thomas Kailath (Stanford, CA, USA)
Peter Lancaster (Calgary, Canada)
Peter D. Lax (New York, NY, USA)
Donald Sarason (Berkeley, CA, USA)
Bernd Silbermann (Chemnitz, Germany)
Harold Widom (Santa Cruz, CA, USA)

B. Malcolm Brown
Michael S.P. Eastham
Karl Michael Schmidt

Periodic Differential Operators

B. Malcolm Brown
Cardiff School of Computer Science &
  Informatics
Cardiff University
Cardiff
United Kingdom

Michael S.P. Eastham
Cardiff School of Computer Science &
  Informatics
Cardiff University
Cardiff
United Kingdom

Karl Michael Schmidt
School of Mathematics
Cardiff University
Cardiff
United Kingdom

ISBN 978-3-0348-0527-8   ISBN 978-3-0348-0528-5 (eBook)
DOI 10.1007/978-3-0348-0528-5
Springer Basel Heidelberg New York Dordrecht London

Library of Congress Control Number: 2012949898

Mathematics Subject Classification (2010): 34A30, 34L05, 34L40, 47A55, 47E05, 81Q10

© Springer Basel 2013

This work is subject to copyright. All rights are reserved by the Publisher, whether the whole or part of the material is concerned, specifically the rights of translation, reprinting, reuse of illustrations, recitation, broadcasting, reproduction on microfilms or in any other physical way, and transmission or information storage and retrieval, electronic adaptation, computer software, or by similar or dissimilar methodology now known or hereafter developed. Exempted from this legal reservation are brief excerpts in connection with reviews or scholarly analysis or material supplied specifically for the purpose of being entered and executed on a computer system, for exclusive use by the purchaser of the work. Duplication of this publication or parts thereof is permitted only under the provisions of the Copyright Law of the Publisher's location, in its current version, and permission for use must always be obtained from Springer. Permissions for use may be obtained through RightsLink at the Copyright Clearance Center. Violations are liable to prosecution under the respective Copyright Law.

The use of general descriptive names, registered names, trademarks, service marks, etc. in this publication does not imply, even in the absence of a specific statement, that such names are exempt from the relevant protective laws and regulations and therefore free for general use.

While the advice and information in this book are believed to be true and accurate at the date of publication, neither the authors nor the editors nor the publisher can accept any legal responsibility for any errors or omissions that may be made. The publisher makes no warranty, express or implied, with respect to the material contained herein.

Printed on acid-free paper

Springer Basel is part of Springer Science+Business Media (www.springer.com)

Contents

Preface   **vii**

1 Floquet Theory   **1**
    1.1 Introduction 1
    1.2 Preliminaries on ordinary differential systems 1
    1.3 Periodic first-order systems 5
    1.4 The discriminant and stability 9
    1.5 Hill's equation and periodic Dirac systems 12
    1.6 Functional properties of Hill's discriminant 15
    1.7 The Mathieu equation 19
    1.8 Periodic, semi-periodic and twisted boundary-value problems . . . 21
    1.9 Appendix: Rofe-Beketov's formula 24
    1.10 Chapter notes 26

2 Oscillations   **31**
    2.1 Introduction 31
    2.2 The Prüfer transform 32
    2.3 The boundary-value problem with separated boundary conditions . 35
    2.4 The rotation number 40
    2.5 Zeros of solutions of Hill's equation 46
    2.6 The upper end-points of the stability intervals 48
    2.7 A step-function example 52
    2.8 Even coefficients 55
    2.9 Comparison of eigenvalues 57
    2.10 Least eigenvalues 60
    2.11 Chapter notes 64

3 Asymptotics   **67**
    3.1 Introduction 67
    3.2 Prüfer transformation formulae 67
    3.3 The coefficient w 70
    3.4 Titchmarsh's asymptotic formula 73

	3.5	Differentiable q	76
	3.6	Length of the instability intervals	79
	3.7	The Mathieu equation	86
	3.8	Asymptotic formulae for solutions	91
	3.9	Absence of instability intervals	94
	3.10	Absence of all but N finite instability intervals	98
	3.11	Absence of odd instability intervals	102
	3.12	All instability intervals non-vanishing	105
	3.13	Chapter notes	108
4	**Spectra**	**113**
	4.1	Introduction	113
	4.2	Regular boundary-value problems	114
	4.3	The spectral function for the half-line problem	120
	4.4	Self-adjoint half-line operators	130
	4.5	The spectrum of the periodic boundary-value problem on the half-line	136
	4.6	The spectral matrix for the full-line problem	142
	4.7	The spectrum of the full-line periodic problem	148
	4.8	Oscillations and spectra	151
	4.9	Bounded solutions and the absolutely continuous spectrum	154
	4.10	Chapter notes	157
5	**Perturbations**	**161**
	5.1	Introduction	161
	5.2	Spectral bands	162
	5.3	Gap eigenvalues	168
	5.4	Critical coupling constants	175
	5.5	Eigenvalue asymptotics	184
	5.6	Chapter notes	193

Bibliography   **197**

Index   **213**

Preface

The work by G.W. Hill on the motion of the lunar perigee, published in Acta Mathematica in 1886, has led to the general name of 'Hill's equation' for the linear second-order ordinary differential equation with periodic coefficients. Already in 1883, G. Floquet laid the foundation of the theory of periodic linear ordinary differential equation systems, and Lyapunov in his 1907 treatise on the stability of motion recognised the fundamental importance of this theory in the context of H. Poincaré's work. Hill's equation gained further eminence with the advent of quantum mechanics in the mid-1920s. The time-independent Schrödinger equation in one spatial dimension with a periodic potential, which may be used to describe the effective electrostatic action on an electron of the regular arrangement of atomic nuclei in a crystal, is of Hill type. Here the spectral parameter of the equation has a direct physical interpretation as the total energy of an electron, and the characteristic pattern of alternating intervals of stability and instability of the equation corresponds to regions of admissible and forbidden energies which, depending on the position of the system's Fermi energy, give an explanation of why some crystals are transparent insulators while others are shiny conductors. Moreover, impurities which perturb the perfect periodic symmetry can lead to additional discrete energy levels in the forbidden regions, thus creating the semiconductors which have had such a pervasive impact on the development of the electronic technology shaping our lives today.

The mathematical study of ordinary differential equation systems with periodic symmetry and of their spectral properties has produced a body of knowledge and a collection of techniques which have been collated in monographs such as W. Magnus and S. Winkler's 1966 book on Hill's equation and, with stronger emphasis on the spectral theory, M. S. P. Eastham's 1973 book which has become a classic on the subject. These spectral properties are also treated as a special case of the general spectral theory of ordinary differential operators in Weidmann's 1987 lecture notes and similar works.

Our motivation for writing the present volume on a subject of such venerable history has arisen from a variety of considerations. Firstly, there has been ongoing progress in the study of periodic differential systems over the past decades, leading to new developments as well as some changes in perspective. For example, the analysis of the higher-dimensional lattice-periodic Schrödinger operator has spun off as a separate subject with its own very specific challenges and connections to other areas such as number theory; see Y. Karpeshina's 1997 monograph [106]. Secondly, there has been a growing interest in the Schrödinger operator's relativistic brother, the Dirac operator. Although some of their spectral properties are very different and many useful techniques for the one-dimensional Schrödinger, or more generally the Sturm-Liouville, operator do not carry over to the Dirac case, due for example to the lower unboundedness of the latter's spectrum, the two differential expressions have much in common when expressed in the form of a Hamiltonian system. From this point of view, the Dirac equation is even the more natural object, whereas in the case of the Sturm-Liouville equation some complications arise from the fact that the weight matrix multiplying the spectral parameter is singular. Finally, the use of oscillation properties of solutions of differential equations has proved to be an effective tool in the study of the spectral properties of the associated differential operators, beginning from C.F. Sturm's observations in the 1830s and H. Prüfer's more powerful reformulation in 1926, the great potential of which was first recognised in J. Weidmann's 1971 paper on oscillation methods for differential equation systems. In the present book, we make systematic use of this technique from an early stage onwards, thus demonstrating that it is not just a trick to count eigenvalues in some specific situations, but that oscillation plays a fundamental and ubiquitous role in the spectral analysis of ordinary differential equation systems of Sturm-Liouville and Dirac type. In fact, it is the close interplay between linear differential equation theory, non-linear oscillation properties and the more abstract theory of linear operators that gives this subject its particular flavour.

In this book we endeavour to give a detailed overview of the techniques and results of the analysis of systems of ordinary differential equations with periodic coefficients, with particular emphasis on the spectral theory of Sturm-Liouville and Dirac operators. It was our aim to provide an introductory text easily accessible to the beginning postgraduate, assuming only elementary knowledge of mathematical analysis, the theory of ordinary differential equations and, in the later parts, of self-adjoint operators in Hilbert space. We decided to include a complete treatment of the singular boundary-value problems on the real line and half-line, following H. Weyl's approach, in as far as it applies to the periodic equation and its perturbations; this includes a number of results which are generally well-known among experts, but for which it is not easy to pinpoint a straightforward proof in the existing textbook literature.

As is natural for a subject of such venerable history, we could not include everything. Periodic equations have given rise to a number of special functions, for example Mathieu and Lamé functions, which have been studied in great detail elsewhere. We therefore feel that in a time when software providing symbolic and numerical computation with these and other special functions is readily available, a circumstantial treatment here would be out of place and consequently we merely refer to the older literature as required. Some aspects of inverse spectral problem as well as extensions such as almost periodic equations or analysis on regular trees are treated only briefly in end-of-chapter notes with literature references. We do, however, include the spectral analysis of perturbed periodic problems, both as a modern development of practical significance and as an instructive application of much of the general theory of the periodic problem explained in this book.

It is a pleasure to thank Prof. Christiane Tretter, Bern, for the encouragement she gave to the idea of this book.

B. Malcolm Brown, Michael S.P. Eastham, Karl Michael Schmidt

Chapter 1

Floquet Theory

1.1 Introduction

The solutions of periodic linear systems of differential equations are not always periodic, but their global qualitative behaviour can be analysed by studying the first period interval only. After a brief summary of the necessary concepts and results from the theory of ordinary differential equations, mainly to introduce the terminology, we establish the existence of Floquet solutions of periodic systems in section 1.3. These are solutions of a particularly simple structure which reveal whether the system is globally stable or unstable. In the case of Hill's equation and periodic Dirac systems, which include a spectral parameter, this classification is determined by a single function called Hill's discriminant, which will be a fundamental tool throughout the book. Its essential properties are studied in sections 1.4 and 1.6. The discriminant function also provides full information about the eigenvalues of periodic and related boundary-value problems on the period interval and its multiples. This connection is explored in section 1.8.

1.2 Preliminaries on ordinary differential systems

The global existence of solutions of the differential equations studied in this book is guaranteed by the following result of ordinary differential equation theory. This result also shows that the solutions depend analytically on a linear parameter in the equation system; this will be important in the study of spectral problems.

Theorem 1.2.1. (a) *Let $F : \mathbb{R} \times \mathbb{C}^n \to \mathbb{C}^n$ be locally integrable w.r.t. the first and locally Lipschitz continuous w.r.t. the second argument, and let*

$$|F(x,v)| \leq g(x)\,|v| + h(x) \qquad (x \in \mathbb{R}, v \in \mathbb{C}^n)$$

with locally integrable g, h, and $u_0 \in \mathbb{C}^n$. Then the initial-value problem

$$u'(x) = F(x, u(x)), \qquad u(0) = u_0$$

has a unique locally absolutely continuous solution $u : \mathbb{R} \to \mathbb{C}^n$.

(b) Assume that $F_1, F_2 : \mathbb{R} \times \mathbb{C}^n \to \mathbb{C}^n$ satisfy the assumptions on F in (a) and are differentiable w.r.t the second argument, with derivatives $\partial_2 F_1, \partial_2 F_2$. Let $u_0 \in \mathbb{C}^n$ and $x_0 \in \mathbb{R}$. Then, denoting by $u(\cdot; \lambda)$ the solution of the initial-value problem

$$u'(x; \lambda) = F_1(x, u(x; \lambda)) + \lambda F_2(x, u(x; \lambda)), \qquad u(0; \lambda) = u_0,$$

$u(x_0; \cdot)$ is an entire function, and $\frac{\partial}{\partial \lambda} u(\cdot; \lambda)$ is the solution of the initial-value problem

$$w'(x) = \partial_2 F_1(x, u(x; \lambda)) w(x) + F_2(x, u(x; \lambda)) + \lambda \partial_2 F_2(x, u(x; \lambda)) w(x),$$
$$w(0; \lambda) = 0.$$

The derivative of a locally absolutely continuous function $u : \mathbb{R} \to \mathbb{C}^n$ is a locally absolutely integrable function u' such that

$$u(x) = u(0) + \int_0^x u' \qquad (x \in \mathbb{R}).$$

The differential equations are thus understood to hold almost everywhere (a.e.).

Let $A : \mathbb{R} \to \mathbb{C}^{n \times n}$ be a locally integrable matrix-valued function. Then, by Theorem 1.2.1 (a), there is a unique locally absolutely continuous solution u of the initial-value problem

$$u' = Au, \quad u(0) = u_0. \tag{1.2.1}$$

The matrix analogue of (1.2.1) is

$$\Psi' = A\Psi, \quad \Psi(0) = \Psi_0, \tag{1.2.2}$$

where $\Psi : \mathbb{R} \to \mathbb{C}^{n \times n}$, and again there is a unique solution Ψ, given Ψ_0, the columns of Ψ being solutions of (1.2.1). As a further consequence of the uniqueness part of Theorem 1.2.1 (a), $\Psi(x)$ is non-singular (or regular) for all x if Ψ_0 is non-singular. In this case Ψ is said to be a *fundamental matrix* for the system

$$u' = Au, \tag{1.2.3}$$

and any solution u of (1.2.3) can be written as

$$u = \Psi c \tag{1.2.4}$$

with a constant vector c, determined uniquely by u.

1.2. Preliminaries on ordinary differential systems

If, further, $\Psi_0 = I$ (the unit $n \times n$ matrix), then Ψ is called the *canonical fundamental matrix* of (1.2.4). Writing $\Psi = \Phi$ in this case, we have

$$\Phi' = A\Phi, \quad \Phi(0) = I. \tag{1.2.5}$$

Any other solution matrix of (1.2.2) can be written as

$$\Psi(x) = \Phi(x)\Psi_0. \tag{1.2.6}$$

We recall two elementary properties of fundamental matrices. The first is Liouville's formula

$$\det \Psi(x) = \det \Psi(0) \exp\left(\int_0^x \operatorname{Tr} A\right). \tag{1.2.7}$$

Here $\det \Psi$ is called the *Wronskian* of Ψ and, by (1.2.7), the Wronskian is constant if $\operatorname{Tr} A = 0$.

The second is the so-called variation of constants formula which relates the solutions of (1.2.3) to those of

$$U' = AU + B, \tag{1.2.8}$$

where B depends on x and it may, or may not, involve U. This formula is given as (1.2.10) in the following proposition.

Proposition 1.2.2. *Let Ψ_i ($i = 1, 2$) satisfy $\Psi_i' = A\Psi_i$, and let $\Psi := \Psi_1 + \Psi_2$ be a fundamental matrix for (1.2.3). Let U be a solution of (1.2.8) and define*

$$U_0(x) = \Psi_1(x)\int_a^x \Psi^{-1}B - \Psi_2(x)\int_x^b \Psi^{-1}B \tag{1.2.9}$$

for fixed points a and b. Then there is a unique constant vector c such that

$$U = \Psi c + U_0. \tag{1.2.10}$$

Proof. By differentiation, we have $U_0' = AU_0 + B$. Hence $(U - U_0)' = A(U - U_0)$, giving $U - U_0 = \Psi c$ by (1.2.4). □

A particular case of (1.2.9) is where, for example, $\Psi_2 = 0$, and then (1.2.9) becomes

$$U_0(x) = \Psi(x)\int_a^x \Psi^{-1}B. \tag{1.2.11}$$

We mention two situations where the inverse Ψ^{-1} in (1.2.9) can be made more explicit. The first is when $n = 2$ with $\Psi_1 = (u\ 0)$ and $\Psi_2 = (0\ v)$. Then it is easy to check that

$$\Psi^{-1} = \begin{pmatrix} -v^T \\ u^T \end{pmatrix} J / \det \Psi,$$

where
$$J = \begin{pmatrix} 0 & 1 \\ -1 & 0 \end{pmatrix}. \tag{1.2.12}$$

Then (1.2.9) becomes
$$U_0(x) = -u(x) \int_a^x \frac{v^T J B}{\det \Psi} - v(x) \int_x^b \frac{u^T J B}{\det \Psi}. \tag{1.2.13}$$

The second situation is where Ψ is the canonical fundamental matrix Φ satisfying (1.2.5) and A has the special form
$$A = JS \tag{1.2.14}$$

with S symmetric and, generalising (1.2.12), $J^T = -J$ and $J^2 = -I$. This time
$$\Phi^{-1} = -J\Phi^T J \tag{1.2.15}$$

and, to see this, consider
$$(J\Phi^T J\Phi)' = J\Phi^T S^T J^T J\Phi + J\Phi^T JJS\Phi = 0$$

by the properties of S and J. Thus
$$J\Phi^T J\Phi = (\text{const.}) = J^2 = -I$$

since $\Phi(0) = I$, and (1.2.15) follows.

The case of a 2×2 system has the special property that, once a non-trivial solution of the equation is known, a second linearly independent solution, and hence a full fundamental system, can be obtained by a quadrature. The form of this second solution is associated with the names of d'Alembert and Rofe-Beketov, and a full account is given in section 1.9. Here we state the particular case of Rofe-Beketov's formula which we require later. We again set $J = \begin{pmatrix} 0 & 1 \\ -1 & 0 \end{pmatrix}$.

Proposition 1.2.3. *Let $X \subset \mathbb{R}$ be an interval, $A : X \to \mathbb{R}^{2\times 2}$, and let $u : X \to \mathbb{R}^2$ be a non-trivial solution of (1.2.3). Further, let $\operatorname{Tr} A = 0$. Then a linearly independent solution v of (1.2.3) is defined by*
$$v(x) = -|u(x)|^{-2} Ju(x) + u(x) \int_{x_0}^x |u|^{-4} u^T (JA - AJ) u$$

where $x_0 \in X$ is fixed.

1.3 Periodic first-order systems

Let $a > 0$ and f a function defined on $\mathbb{R}$. Then we call f a-*periodic* if
$$f(x+a) = f(x) \qquad (x \in \mathbb{R}).$$

The function f is called a-*semi-periodic* if
$$f(x+a) = -f(x) \qquad (x \in \mathbb{R});$$

clearly any a-semi-periodic function is $2a$-periodic.

Now let $n \in \mathbb{N}$ and consider a locally integrable matrix-valued a-periodic function $A : \mathbb{R} \to \mathbb{C}^{n \times n}$. The analysis of the periodic system
$$u' = A u \tag{1.3.1}$$

is based upon the following fundamental observation.

Theorem 1.3.1. *Let $A : \mathbb{R} \to \mathbb{C}^{n \times n}$ be a-periodic and $\Phi : \mathbb{R} \to \mathbb{C}^{n \times n}$ the canonical fundamental matrix of (1.3.1). Then there is a unique matrix $M \in \mathbb{C}^{n \times n}$, called* the monodromy matrix*, such that*
$$\Phi(x+a) = \Phi(x) M \qquad (x \in \mathbb{R}). \tag{1.3.2}$$

In fact,
$$M = \Phi(a). \tag{1.3.3}$$

Proof. Define $\Psi(x) = \Phi(x + a)$. Then $\Psi'(x) = A(x+a)\Phi(x+a) = A(x)\Psi(x)$. Thus Ψ is a matrix solution of the first-order system and can be expressed in terms of the canonical fundamental matrix as in (1.2.6), giving
$$\Psi(x) = \Phi(x)\Psi(0) = \Phi(x)\Phi(a).$$

Thus (1.3.2) holds and we have the existence of a monodromy matrix satisfying (1.3.2) and (1.3.3). Its uniqueness follows from the pointwise regularity of Φ. □

On the basis of this theorem, one can study the periodic system by investigating the properties of the matrix M, in particular its eigenvalues. Since M is a regular matrix, all its eigenvalues are non-zero and hence can be represented in the form e^μ, where $\mu \in \mathbb{C}$.

Theorem 1.3.2. *Let M be the monodromy matrix of an a-periodic linear system (1.3.1) and let $\mu \in \mathbb{C}$. Then e^μ is an eigenvalue of M if and only if there is a non-trivial solution $u : \mathbb{R} \to \mathbb{C}^n$ such that*
$$u(x+a) = e^\mu u(x) \qquad (x \in \mathbb{R}). \tag{1.3.4}$$

Proof. Let $v \in \mathbb{C}^n \setminus \{0\}$ be an eigenvector of M for eigenvalue e^μ, and set $u := \Phi v$, where Φ is the canonical fundamental matrix. Then u is a solution, and from (1.3.2) we conclude that

$$u(x+a) = \Phi(x+a)\,v = \Phi(x)\,M\,v = e^\mu\,\Phi(x)\,v = e^\mu\,u(x) \qquad (x \in \mathbb{R}).$$

Conversely, if $u : \mathbb{R} \to \mathbb{C}^n$ is a non-trivial solution satisfying (1.3.4), then $u(0) \neq 0$ and

$$M\,u(0) = \Phi(a)\,u(0) = u(a) = e^\mu\,u(0).$$

Thus $u(0)$ is an eigenvector of M. $\square$

The eigenvalues of M are called *Floquet multipliers*, the corresponding values μ are called *Floquet exponents*, and any non-trivial solution satisfying (1.3.4) is called a *Floquet solution* of the periodic linear system. If u is a Floquet solution with Floquet exponent μ, then

$$u(x) = e^{\mu x/a}\,g(x), \qquad (x \in \mathbb{R}) \tag{1.3.5}$$

where the function $g(x) = e^{-\mu x/a}\,u(x)$ is a-periodic. Thus, a Floquet solution is always a product of a real or complex exponential and an a-periodic function.

If all eigenvalues of M have geometric multiplicity equal to their algebraic multiplicity, then there are n linearly independent eigenvectors and correspondingly n linearly independent Floquet solutions. Then all solutions of the system can be expressed as linear combinations of Floquet solutions.

In general, however, M has distinct eigenvalues $\exp \mu_k$ ($1 \leq k \leq l$), which are simple, and $\exp \mu_{l+K}$ ($1 \leq K \leq L$), which have algebraic multiplicity r_K (≥ 2). To obtain the representation of solutions which now corresponds to (1.3.5), we introduce the matrix $R \in \mathbb{C}^{n \times n}$ defined by

$$M = \exp aR, \tag{1.3.6}$$

such an R existing because M is regular. A first consequence of (1.3.6) is that the matrix-valued function $H(x) = \Phi(x)\exp(-xR)$ is a-periodic since, by (1.3.2),

$$H(x+a) = \Phi(x)M\exp(-aR)\exp(-xR) = H(x).$$

Thus $\Phi(x)$ has the representation

$$\Phi(x) = H(x)\exp xR. \tag{1.3.7}$$

Next, to make (1.3.7) more explicit, we introduce the Jordan normal form of R:

$$R = TBT^{-1} \tag{1.3.8}$$

where T is non-singular and B is the diagonal partitioned matrix

$$B = \mathrm{dg}\,(B_0, B_1, \ldots, B_L)$$

1.3. Periodic first-order systems

in which
$$B_0 = \text{dg}\,(\lambda_1, \ldots, \lambda_l),$$
B_K $(1 \leq K \leq L)$ is the $r_K \times r_K$ matrix

$$\begin{pmatrix} \lambda_{l+K} & 1 & 0 & \cdots & 0 \\ 0 & \lambda_{l+K} & 1 & & \\ \vdots & & \ddots & & \\ \vdots & & & \ddots & \\ 0 & & & 0 & \lambda_{l+K} \end{pmatrix}$$

and
$$a\lambda_k = \mu_k, \quad a\lambda_{l+K} = \mu_{l+K}.$$

This nature of B_0 and the B_K arises from (1.3.6) and (1.3.8) since

$$M = T(\exp aB)T^{-1}$$

and

$$\exp aB = \text{dg}\,(\exp aB_0, \exp aB_1, \ldots, \exp aB_L) \tag{1.3.9}$$

where

$$\exp aB_0 = \text{dg}\,(\exp a\lambda_1, \ldots, \exp a\lambda_l)$$

and

$$\exp aB_K = (\exp a\lambda_{l+K}) \begin{pmatrix} 1 & a & \cdots & a^{r_K-1}/(r_K-1)! \\ 0 & 1 & a & \cdots a^{r_K-2}/(r_K-2)! \\ \vdots & & & \vdots \\ 0 & \cdots & 0 & 1 \end{pmatrix}.$$

With T as in (1.3.8), we define the fundamental matrix Ψ by $\Psi = \Phi T$. Then, by (1.3.7) and (1.3.8),

$$\Psi(x) = H(x)(\exp xR)T = H(x)TT^{-1}(\exp xR)T$$
$$= G(x)\exp xB \tag{1.3.10}$$

where $G(x)$ is a-periodic. Now, finally, we denote the column vectors of Ψ and G by ψ_i and g_i respectively, and note that $\exp xB$ is given by (1.3.9) with x in place of a. Then from (1.3.10) we obtain the representation of n linearly independent solutions of (1.3.1) in the form

$$\psi_k(x) = (\exp \mu_k x/a)g_k(x) \quad (1 \leq k \leq l),$$

$$\psi_{l+K+s}(x) = (\exp \mu_{l+K} x/a) \sum_{t=0}^{s} \frac{x^t}{t!} g_{l+K+s-t}(x) \quad (0 \leq s \leq r_K - 1,\ 1 \leq K \leq L),$$

$$\tag{1.3.11}$$

where the g_i are a-periodic.

As an example of Theorems 1.3.1 and 1.3.2, consider

$$A = \begin{pmatrix} 0 & 1 \\ -q^2 & q'/q \end{pmatrix} \tag{1.3.12}$$

in (1.3.1), where q is nowhere zero. Then it is easy to check that

$$\Phi = \begin{pmatrix} \cos Q & q(0)^{-1} \sin Q \\ -q \sin Q & q(0)^{-1} q \cos Q \end{pmatrix}$$

where $Q(x) = \int_0^x q$. Then, since $q(a) = q(0)$, the eigenvalues of M $(= \Phi(a))$ are $\exp(\pm i Q(a))$. Thus the Floquet exponents are $\pm i Q(a)$.

Theorem 1.3.2 shows in particular that an a-periodic system has an a-periodic solution if and only if 1 is an eigenvalue of its monodromy matrix M. This will not usually be the case, and for the equations we shall study later it is the exception rather than the rule. There is nevertheless a special class of periodic linear systems for which *all* solutions are periodic with the same period, as follows.

Theorem 1.3.3. *Let $A : \mathbb{R} \to \mathbb{C}^{n \times n}$ be a-periodic and assume that there is a continuously differentiable function $h : [-1, 1] \to \mathbb{R}$ such that*

$$h(1) - h(-1) = a \tag{1.3.13}$$

and, with $F(s) = \int_0^s A$ ($s \in \mathbb{R}$), $F \circ h$ is an even function. Then all solutions of $u' = Au$ have period a.

Proof. Let Φ be the canonical fundamental matrix of $u' = Au$. Then $Y := \Phi \circ h$ satisfies

$$Y' = h' \Phi' \circ h = h' (A \circ h)(\Phi \circ h) = (F \circ h)' Y.$$

Also, $Z(t) := Y(-t)$ $(t \in [-1, 1])$ has $Z(0) = Y(0)$ and

$$Z'(t) = -Y'(-t) = (F \circ h)'(t) Z(t),$$

because $(F \circ h)'$ is odd. By the uniqueness of the solution of initial-value problems, $Z = Y$. Thus Y is even and hence, by (1.3.13) and (1.3.2),

$$\Phi(h(-1)) = Y(-1) = Y(1) = \Phi(h(1)) = \Phi(h(-1)) M.$$

Since Φ is pointwise regular, this implies that $M = I$. The theorem now follows from (1.3.2). $\square$

The hypotheses of Theorem 1.3.3 are satisfied if for example A is an odd function itself; then one can choose $h(t) = \frac{1}{2} a t$ ($t \in [-1, 1]$).

1.4 The discriminant and stability

From now on we concentrate on the fertile theory for the case of a 2×2 system

$$u' = Au \tag{1.4.1}$$

where A has the structure

$$A = JS \tag{1.4.2}$$

with

$$J = \begin{pmatrix} 0 & 1 \\ -1 & 0 \end{pmatrix} \tag{1.4.3}$$

and S is a real a-periodic symmetric 2×2 matrix function. It follows that $\operatorname{Tr} A = 0$ and hence, as already noted, Liouville's formula (1.2.7) implies that $\det \Psi$ is constant for any solution matrix Ψ. In particular (1.3.3) gives

$$\det M = \det \Phi(a) = \det \Phi(0) = 1. \tag{1.4.4}$$

The Floquet multipliers are the eigenvalues ρ_1 and ρ_2 of M and hence, by (1.4.4), they are the solutions of the quadratic equation

$$\rho^2 - (\operatorname{Tr} M)\rho + 1 = 0.$$

Thus the product $\rho_1 \rho_2 = 1$ and hence, as in section 1.3,

$$\rho_1 = e^\mu, \quad \rho_2 = e^{-\mu} \tag{1.4.5}$$

for some $\mu \in \mathbb{C}$. Since M is real, ρ_1 and ρ_2 are either both real with the same sign, or complex conjugates with modulus 1.

The position of the Floquet multipliers in the complex plane, and hence the nature of the solutions of (1.4.1), can be classified by noting that the sum $\rho_1 + \rho_2$ is equal to $\operatorname{Tr} M$. The real number D defined by

$$D := \operatorname{Tr} M = 2 \cosh \mu \tag{1.4.6}$$

is called the *Hill discriminant* of the periodic system (1.4.1), and there are the following five cases distinguished by the value of D.

Case 1. $D > 2$. Here $\mu \in \mathbb{R} \setminus \{0\}$ in (1.4.5) and (1.4.6). By Theorem 1.3.2 and (1.3.5) there exist Floquet solutions $u_\pm$ of the form

$$u_\pm(x) = e^{\pm \mu x / a} g_\pm(x) \quad (x \in \mathbb{R}) \tag{1.4.7}$$

with a-periodic $g_\pm$. Since one of the Floquet solutions has an exponentially growing envelope, and the other an exponentially decreasing envelope, they are linearly independent and form a fundamental system. Hence every non-trivial solution of the periodic equation is unbounded.

Case 2. $D < -2$. The situation here is the same as in Case 1 except that the Floquet multipliers are now negative. Hence now $\mu \in \mathbb{R}\setminus\{0\} + \pi i$.

Case 3. $|D| < 2$. Here $\mu = i\nu$ with $\nu \in \mathbb{R}\setminus\pi\mathbb{Z}$ and $|D| = 2|\cos\nu|$. This is the case of non-real Floquet multipliers, being complex conjugates. By Theorem 1.3.2 and (1.3.5) there exists a Floquet solution u of the form

$$u(x) = e^{i\nu x/a} g(x) \qquad (x \in \mathbb{R}) \qquad (1.4.8)$$

with a-periodic g. Its complex conjugate $\bar{u}$ is a Floquet solution for the other Floquet multiplier, and is linearly independent of u; indeed, if there were a constant $c \in \mathbb{C}\setminus\{0\}$ such that $\bar{u} = cu$, then

$$0 = \overline{u(a)} - \overline{u(a)} = cu(a) - e^{-i\nu}\overline{u(0)} = (e^{i\nu} - e^{-i\nu})cu(0),$$

which is excluded in the current case. In view of (1.4.8), the solutions u, $\bar{u}$ are globally bounded and, as they form a fundamental system, so also is every other solution.

Case 4. $D = 2$. Here $\mu = 0$ in (1.4.6) and (1.4.5) and the Floquet multipliers are both 1. Thus the monodromy matrix M has an algebraically double eigenvalue. If the geometric multiplicity is 2, then $M = I$, and so, by Theorem 1.3.1, all solutions of (1.4.1) are a-periodic. In this case periodic solutions are said to *coexist*.

If, on the other hand, the geometric multiplicity is only 1, there is still an a-periodic Floquet solution u, but there are no linearly independent Floquet solutions. This is the case $l = 0$, $L = K = 1$, $r_K = 2$ of (1.3.11). Here, however, there is a simpler way of obtaining a second solution v by using Rofe-Beketov's formula (Proposition 1.2.3) which is applicable since $\operatorname{Tr} A = 0$. Thus

$$v(x) = -|u(x)|^{-2} Ju(x) + u(x) \int_{x_0}^{x} |u|^{-4} u^T (JA - AJ) u$$
$$= G(x) + u(x) \int_{0}^{x} F$$

say, where G and F are a-periodic. If we now define

$$H(x) = \int_{0}^{x} F - \frac{x}{a}\int_{0}^{a} F,$$

then H is a-periodic and we can write v in the form

$$v(x) = g(x) + cxu(x) \qquad (x \in \mathbb{R}), \qquad (1.4.9)$$

where g is a-periodic and c is a constant. Clearly $c \neq 0$, as otherwise v would be a-periodic, leading to coexistence as above. Therefore all solutions except for the multiples of u are unbounded.

Case 5. $D = -2$. Here $\mu = \pi i$ in (1.4.6) and (1.4.5) and the Floquet multipliers

1.4. The discriminant and stability

are both -1. The situation now is similar to that in Case 4, the only difference being that u and g are a-semi-periodic.

We now introduce the following definition concerning the boundedness of solutions of (1.4.1).

Definition 1.4.1. The system (1.4.1) is said to be (a) *unstable* if all non-trivial solutions are unbounded in $\mathbb{R}$, (b) *conditionally stable* if there is a non-trivial solution which is bounded in $\mathbb{R}$, (c) *stable* if all solutions are bounded in $\mathbb{R}$.

The above classification shows that (1.4.1) is unstable if $|D| > 2$ and stable if $|D| < 2$. When $D = \pm 2$, it is also stable if $M = \pm I$ respectively, and otherwise conditionally stable. We also now state explicitly the implications of the classification for the existence of periodic and semi-periodic solutions.

Theorem 1.4.2. *The system (1.4.1) has non-trivial a-periodic solutions if and only if $D = 2$, and a-semi-periodic solutions if and only if $D = -2$. All solutions of (1.4.1) are a-periodic or a-semi-periodic if and only if, in addition, $M = \pm I$ respectively.*

There is one further case to mention to cover situations where λ may be complex-valued.

Case 6. D non-real. By (1.4.6), μ is non-real with $\operatorname{Re} \mu \neq 0$. Also, (1.4.7) continues to hold and hence all non-trivial solutions are unbounded, making (1.4.1) unstable.

Example 1. For the first of two examples on the above theory, let A in (1.4.1) have the form

$$A = \begin{pmatrix} 0 & 1 \\ -q & 0 \end{pmatrix}$$

where q is a step-function:

$$q = q_i^2 \;(> 0) \text{ on } X_i := ((i-1)a/n, ia/n] \;\; (i = 1, \ldots, n),$$

and the sub-intervals X_i are of equal length $1/n$ for convenience in this example. Then (1.4.1) is equivalent to the second-order equation

$$y'' + qy = 0, \qquad (1.4.10)$$

where y is the first component of u. An equation of the type (1.4.10) with a step-function coefficient is known as a *Meissner equation*. If we introduce the matrix

$$Q_i = \begin{pmatrix} \cos(aq_i/n) & q_i^{-1}\sin(aq_i/n) \\ -q_i \sin(aq_i/n) & \cos(aq_i/n) \end{pmatrix} \qquad (1.4.11)$$

then it is easy to check that the monodromy matrix M is given by the product $M = Q_n Q_{n-1} \ldots Q_1$.

Also, the change of variable x to $-x$ (i.e., a reflection) in (1.4.10) gives another equation of the same type but with the two Floquet multipliers interchanged in (1.4.5). The new monodromy matrix M_1 is $M_1 = Q_1 Q_2 \ldots Q_n$, and $\operatorname{Tr} M_1 = \operatorname{Tr} M$. Thus, by (1.4.6), the stability nature of (1.4.10) is unchanged if the q_i are permuted by cyclic change or reflection (or a combination of these).

Other permutations, however, can change the stability nature. Thus, taking $n = 4$, $a = 2\pi$, $q_1 = q_2 = c$ and $q_3 = q_4 = 1$, we can consider the permutations represented by the monodromy matrices $M = Q_1 Q_2 Q_1 Q_2$ and $M' = Q_1^2 Q_2^2$, where

$$Q_1 = \begin{pmatrix} \cos(c\pi/2) & c^{-1}\sin(c\pi/2) \\ -c\sin(c\pi/2) & \cos(c\pi/2) \end{pmatrix}, \quad Q_2 = \begin{pmatrix} 0 & 1 \\ -1 & 0 \end{pmatrix}$$

as in (1.4.11). Then

$$\operatorname{Tr} M = (c^2 + c^{-2})\sin^2(c\pi/2) - 2\cos^2(c\pi/2), \qquad \operatorname{Tr} M' = -2\cos c\pi.$$

The choice $c = 5/2$ gives $\operatorname{Tr} M = \frac{17}{8} + \frac{2}{25} > 2$ (instability), and $\operatorname{Tr} M' = 0$ (stability).

Example 2. A simple, but non-trivial example where an explicit Floquet multiplier ρ appears is provided by (1.4.10) but with a different choice of q as follows. Consider the function y defined by

$$y(x) = \exp\left(\int_0^x P(t) \cos t \, dt\right) \cos x,$$

where P is 2π–periodic and absolutely continuous. Then differentiation shows that y is a solution of

$$y''(x) + (1 + 3P(x)\sin x - P'(x)\cos x - P^2(x)\cos^2 x)\, y(x) = 0$$

and, further, $y(x + 2\pi) = \rho y(x)$ with

$$\rho = \exp\left(\int_0^{2\pi} P(t) \cos t \, dt\right).$$

Thus $\mu = \int_0^{2\pi} P(t) \cos t \, dt$ in (1.3.4).

1.5 Hill's equation and periodic Dirac systems

The theory of the previous section takes on a new significance when S involves a real parameter λ in the form $S = B + \lambda W$, where B and W are symmetric with W positive semidefinite and $W \neq 0$. Thus, by (1.4.2), the system $u' = Au$ is now

$$u' = J(B + \lambda W)u. \tag{1.5.1}$$

1.5. Hill's equation and periodic Dirac systems

We also assume that W has the property that no non-trivial solution u of (1.4.1) has $Wu = 0$ a.e. on an open interval: otherwise there would be non-trivial solutions which are independent of λ, rendering the inclusion of the spectral parameter ineffective. Although λ is taken to be real here, it is sometimes necessary to allow it to be complex, and we shall comment on this as the occasion arises. We shall see that the role of λ is as a spectral parameter.

The general properties of the discriminant D, as a function of λ, is the subject of the next section. First however, we note that the framework (1.5.1) covers two types of equation which play a central role in mathematical physics. The first is the periodic Sturm-Liouville, or Hill's, equation

$$-(py')' + qy = \lambda w y, \qquad (1.5.2)$$

where $1/p, q, w$ are real-valued, locally integrable, a-periodic functions, with $p > 0$ and $w > 0$. This equation occurs in many physical problems. For example, with constant p and w, this is the one-dimensional Schrödinger equation of non-relativistic quantum mechanics. As $-i\frac{d}{dx}$ is the quantum mechanical momentum operator, the first term represents the kinetic energy, q is the potential and λ is the total energy parameter, and so (1.5.2) expresses an energy balance. The same type of equation arises in the modelling of a vibrating string; here $q = 0$, and the first term, with p determined by the local mass and stiffness of the string, represents elastic energy, while λ is the square of the angular frequency of oscillation in time.

The second-order equation (1.5.2) is equivalent to the periodic 2×2 system for the vector-valued function

$$u = \begin{pmatrix} y \\ py' \end{pmatrix} \qquad (1.5.3)$$

with coefficient matrix

$$A = \begin{pmatrix} 0 & 1/p \\ q - \lambda w & 0 \end{pmatrix} = \begin{pmatrix} 0 & 1 \\ -1 & 0 \end{pmatrix} \left(\begin{pmatrix} -q & 0 \\ 0 & 1/p \end{pmatrix} + \lambda \begin{pmatrix} w & 0 \\ 0 & 0 \end{pmatrix} \right)$$

of the form in (1.5.1). We call this first-order system the *Sturm-Liouville system*. The combination py', which will be locally absolutely continuous for solutions y while the derivative y' may not be, is called the *quasiderivative* of y. Note that, if $Wu = 0$ on an open interval for a solution u of the Sturm-Liouville system, then $u_1 = y = 0$ on that interval and hence $py' = 0$ too, and so by the uniqueness theorem u must be trivial.

The second equation is the periodic Dirac system

$$-i\sigma_2 u' + p_1 \sigma_3 u + p_2 \sigma_1 u + qu = \lambda u, \qquad (1.5.4)$$

where p_1, p_2, q are real-valued locally integrable, a-periodic functions and

$$\sigma_1 = \begin{pmatrix} 0 & 1 \\ 1 & 0 \end{pmatrix}, \quad \sigma_2 = \begin{pmatrix} 0 & -i \\ i & 0 \end{pmatrix}, \quad \sigma_3 = \begin{pmatrix} 1 & 0 \\ 0 & -1 \end{pmatrix}$$

are the Pauli matrices, which satisfy the algebraic relations

$$\sigma_j \sigma_k + \sigma_k \sigma_j = 2\delta_{jk} I \qquad (j, k \in \{1, 2, 3\}).$$

The equation (1.5.4) arises in the relativistic quantum mechanics of a particle in one dimension. Here q is the potential and λ the energy parameter as for the Schrödinger equation. The first term is the quantum-mechanical momentum operator multiplied by the matrix σ_2; with constant particle mass p_1 one finds that

$$\left(-i\sigma_2 \frac{d}{dx} + p_1 \sigma_3\right)^2 = \left(-\frac{d^2}{dx^2} + p_1^2\right) I,$$

and so the first two terms of the differential operator of (1.5.4) are in a sense a square root of the first term in (1.5.2) plus a constant.

Here the corresponding matrix A has the form

$$A = \begin{pmatrix} -p_2 & p_1 - q + \lambda \\ p_1 + q - \lambda & p_2 \end{pmatrix} = \begin{pmatrix} 0 & 1 \\ -1 & 0 \end{pmatrix} \left(\begin{pmatrix} -p_1 - q & -p_2 \\ -p_2 & p_1 - q \end{pmatrix} + \lambda \begin{pmatrix} 1 & 0 \\ 0 & 1 \end{pmatrix} \right),$$

again as in (1.5.1).

Both equations (1.5.2) and (1.5.4) occur in the quantum mechanics of a one-dimensional electron moving in a periodic force field, such as the effective electrostatic field of the periodic array of atomic nuclei in a crystal. The non-relativistic Schrödinger equation is of the type of Hill's equation, whereas the Dirac system gives a relativistic description. In both cases, the spectral parameter λ represents the particle energy.

Finally in this section, we return to the Sturm-Liouville equation (1.5.2) and give the explicit form of the discriminant D which arises from the use of (1.5.3). Let ϕ_1 and ϕ_2 be the solutions of (1.5.2) which satisfy the initial conditions

$$\phi_1(0) = 1, \; \phi_1'(0) = 0; \quad \phi_2(0) = 0, \; \phi_2'(0) = 1. \tag{1.5.5}$$

Then, by (1.5.3) and (1.2.5),

$$\Phi(x) = \begin{pmatrix} 1 & 0 \\ 0 & \frac{1}{p(0)} \end{pmatrix} \begin{pmatrix} \phi_1 & \phi_2 \\ p\phi_1' & p\phi_2' \end{pmatrix}(x).$$

Since $p(0) = p(a)$, (1.3.3) gives

$$M = \begin{pmatrix} \phi_1 & \phi_2 \\ \phi_1' & \phi_2' \end{pmatrix}(a)$$

and hence, by (1.4.6),

$$D = \operatorname{Tr} M = \phi_1(a) + \phi_2'(a). \tag{1.5.6}$$

1.6 Functional properties of Hill's discriminant

We return now to the general periodic system (1.5.1) and study the properties of Hill's discriminant D as a function of λ, under the stated conditions on B and W. Primarily, λ is a real parameter and, consequently, D is real-valued and continuous. We do however observe in passing that, if λ is allowed to be complex, then D is an analytic function of λ.

We introduce the sets

$$\mathcal{S} = \{\lambda \in \mathbb{R} \mid |D(\lambda)| < 2\}, \quad \mathcal{I} = \{\lambda \in \mathbb{R} \mid |D(\lambda)| > 2\}$$

which, as open subsets of $\mathbb{R}$, consist of at most countably many open intervals. In view of Definition 1.4.1, these intervals are called the *stability intervals* and *instability intervals*, respectively.

Moreover, the complementary sets $\{\lambda \in \mathbb{R} \mid D(\lambda) = 2\}$ and $\{\lambda \in \mathbb{R} \mid D(\lambda) = -2\}$ consist of at most countably many isolated points. This follows from the following properties of the derivatives of D, in which M again denotes the monodromy matrix (1.3.3).

Theorem 1.6.1. (a) *On each subinterval of $\mathcal{S}$, the function D is strictly monotonic.*

(b) *At a point $\lambda \in \mathbb{R}$ where $|D(\lambda)| = 2$, $D'(\lambda) = 0$ if and only if*

$$M(\lambda) = (\operatorname{sgn} D(\lambda))\, I; \tag{1.6.1}$$

in this case $D(\lambda)\, D''(\lambda) < 0$.

Proof. (a) We begin by noting the inequality

$$\left| t + \frac{1}{t} \right| \geq 2 \quad (t \in \mathbb{R}\setminus\{0\}) \tag{1.6.2}$$

with equality only if $|t| = 1$. This will be required later in the proof, and it follows from the identity

$$\left(t + \frac{1}{t}\right)^2 = \left(t - \frac{1}{t}\right)^2 + 4.$$

Now let Φ be the canonical fundamental matrix of (1.5.1),

$$\Phi' = J(B + \lambda W)\,\Phi, \quad \Phi(0) = I.$$

Then, for each $x \in \mathbb{R}$, $\Phi(x, \lambda)$ is a differentiable function of λ and its λ-derivative satisfies the formally differentiated initial-value problem

$$\frac{d}{dx}\left(\frac{\partial \Phi}{\partial \lambda}\right) = J(B + \lambda W)\frac{\partial \Phi}{\partial \lambda} + JW\Phi, \quad \frac{\partial \Phi(0, \lambda)}{\partial \lambda} = 0. \tag{1.6.3}$$

By the method of variation of constants (see (1.2.10) and (1.2.11)), $\partial \Phi/\partial \lambda$ can be expressed in terms of Φ as

$$\frac{\partial \Phi(x,\lambda)}{\partial \lambda} = \Phi(x,\lambda) \int_0^x \Phi^{-1} J W \Phi. \tag{1.6.4}$$

Thus

$$D'(\lambda) = \operatorname{Tr} \frac{\partial \Phi(a,\lambda)}{\partial \lambda} = \operatorname{Tr}\left(\Phi(a,\lambda) \int_0^a \Phi^{-1} J W \Phi\right) \tag{1.6.5}$$

where, denoting the columns of Φ by u and v, the integrand can be written in the form

$$\Phi^{-1} J W \Phi = \begin{pmatrix} v^T W u & v^T W v \\ -u^T W u & -u^T W v \end{pmatrix}. \tag{1.6.6}$$

Now $D = \operatorname{Tr} \Phi(a) = u_1(a) + v_2(a)$ where, for economy, we no longer exhibit the dependence on λ. Then, since $\det \Phi(a) = 1$, we obtain

$$D^2 = (u_1(a) - v_2(a))^2 + 4u_1(a)v_2(a) = (u_1(a) - v_2(a))^2 + 4 + 4u_2(a)v_1(a).$$

Combining this with (1.6.5) and observing that $u^T W v = v^T W u$, we obtain

$$4v_1(a) D' = (D^2 - 4) \int_0^a v^T W v \tag{1.6.7}$$
$$- \int_0^a [(u_1(a) - v_2(a))v - 2v_1(a)u]^T W [(u_1(a) - v_2(a))v - 2v_1(a)u].$$

Since W is positive semidefinite, it follows that

$$4v_1(a) D' \leq (D^2 - 4) \int_0^a v^T W v.$$

For the same reason, $\int_0^a v^T W v \geq 0$ as well; in fact, it is strictly positive because, if $\int_0^a v^T W v = 0$, then $W v = \sqrt{W}\sqrt{W} v = 0$ a.e. on $[0,a]$, contradicting our assumptions on W and the fact that v is a non-trivial solution.

Moreover, $v_1(a) \neq 0$ if λ lies in $\mathcal{S}$. Indeed, if $v_1(a) = 0$, then

$$u_1(a) v_2(a) = \det \Phi(a) = 1,$$

and so

$$|D| = |\operatorname{Tr} \Phi(a)| = \left|u_1(a) + \frac{1}{u_1(a)}\right| \geq 2$$

by (1.6.2).

Hence we conclude that $D' \neq 0$ on $\mathcal{S}$, and so D is strictly monotonic on each subinterval of $\mathcal{S}$, increasing if $v_1(a) < 0$ and decreasing if $v_1(a) > 0$.

(b) If $\Phi(a, \lambda) = M(\lambda) = \pm I$, then $|D(\lambda)| = 2$ and it is clear from (1.6.5) that

$$D'(\lambda) = 0. \tag{1.6.8}$$

1.6. Functional properties of Hill's discriminant

Conversely, if $|D(\lambda)| = 2$ and $D'(\lambda) = 0$, then (1.6.7) shows that

$$\int_0^a [(u_1(a) - v_2(a))v - 2v_1(a)u]^T W [(u_1(a) - v_2(a))v - 2v_1(a)u] = 0,$$

which implies that $(u_1(a) - v_2(a))v - 2v_1(a)u$ is the trivial solution. Since u and v are linearly independent, we conclude that $v_1(a) = 0$ and $u_1(a) = v_2(a) = \operatorname{sgn} D(\lambda)$. Using this information about the matrix $\Phi(a, \lambda)$ in (1.6.5), we find that

$$u_2(a) \int_0^a v^T W v = D'(\lambda) = 0.$$

Thus $u_2(a) = 0$ and $M(\lambda) = (\operatorname{sgn} D(\lambda))I$.

The second derivative of Φ with respect to λ is the solution of the formally differentiated initial-value problem (1.6.3),

$$\frac{d}{dx}\left(\frac{\partial^2 \Phi}{\partial \lambda^2}\right) = J(B + \lambda W)\frac{\partial^2 \Phi}{\partial \lambda^2} + 2JW\frac{\partial \Phi}{\partial \lambda}, \quad \frac{\partial^2 \Phi(0, \lambda)}{\partial \lambda^2} = 0.$$

Solving by variation of constants and inserting the integral representation (1.6.4) for the first derivative, we obtain

$$D''(\lambda) = \operatorname{Tr} \frac{\partial^2 \Phi(a, \lambda)}{\partial \lambda^2} = \operatorname{Tr} \left(2\Phi(a, \lambda) \int_0^a F(t) \int_0^t F(s)\, ds\, dt \right),$$

where

$$F = \Phi^{-1} JW \Phi. \tag{1.6.9}$$

Since $\Phi(a, \lambda) = (\operatorname{sgn} D(\lambda))I$ and $\operatorname{Tr}(F(t)F(s)) = \operatorname{Tr}(F(s)F(t))$ ($s, t \in [0, a]$), we obtain

$$\frac{1}{2}(\operatorname{sgn} D(\lambda))D''(\lambda) = \int_0^a \int_0^t \operatorname{Tr}(F(t)F(s))\, ds\, dt$$

$$= \frac{1}{2}\int_0^a \int_0^a \operatorname{Tr}(F(t)F(s))\, ds\, dt = \frac{1}{2}\operatorname{Tr}\left(\int_0^a F\right)^2$$

$$= \left(\int_0^a u^T W v\right)^2 - \left(\int_0^a v^T W v\right)\left(\int_0^a u^T W u\right) \tag{1.6.10}$$

by (1.6.6) and (1.6.9). Hence $(\operatorname{sgn} D(\lambda))D''(\lambda) \leq 0$ by the Cauchy-Schwarz inequality. The inequality is strict, since u and v are linearly independent. $\square$

Theorem 1.6.1 gives the following picture of the general behaviour of Hill's discriminant on the real line. Inside a stability interval, D is strictly monotonic increasing or decreasing with range $(-2, 2)$. As an end-point of a stability interval is reached, either D crosses ± 2 and an instability interval ensues, or D just touches ± 2 and turns back, producing an immediately adjacent stability interval. In the first case, the point where $|D| = 2$ is a point of conditional stability, in the second case it is a point of coexistence. Since the first derivative of D vanishes at a point of coexistence, but the second derivative is non-zero, points of coexistence are always isolated.

We call the connected components of $\mathbb{R} \setminus S$ the *closed instability intervals*. If a closed instability interval has positive length, it is indeed the closure of an instability interval; otherwise it consists of a single coexistence point separating two stability intervals.

Equation (1.6.5) expresses the derivative of the discriminant in terms of the canonical fundamental system Φ. For spectral parameter λ inside a stability interval, this gives rise to a suggestive formula for the derivative of the discriminant in terms of the complex Floquet solution (1.4.8). Indeed, let $\lambda \in S$, and let $Y(x, \lambda) = (u(x, \lambda), \overline{u}(x, \lambda))$ be the fundamental matrix composed of the Floquet solution $u(x, \lambda)$ and its complex conjugate, where

$$u(x + a, \lambda) = e^{i\nu} u(x, \lambda) \qquad (x \in \mathbb{R})$$

and $\nu \in \mathbb{R}$. Then $Y(0, \lambda)$ is a matrix of eigenvectors of the monodromy matrix $M(\lambda)$, and in particular

$$Y(0, \lambda)^{-1} \Phi(a, \lambda) Y(0, \lambda) = \begin{pmatrix} e^{i\nu} & 0 \\ 0 & e^{-i\nu} \end{pmatrix}. \tag{1.6.11}$$

Furthermore, $Y(x, \lambda) = \Phi(x, \lambda) Y(0, \lambda)$, and so $\Phi(x, \lambda) = Y(x, \lambda) Y(0, \lambda)^{-1}$. From (1.6.5) and (1.6.11) we thus obtain, suppressing the variable λ for convenience,

$$\begin{aligned} D'(\lambda) &= \mathrm{Tr}\left(\Phi(a) \int_0^a Y(0) Y(t)^{-1} JW Y(t) Y(0)^{-1} dt\right) \\ &= \mathrm{Tr}\left(Y(0)^{-1} \Phi(a) Y(0) \int_0^a Y^{-1} JW Y\right) \\ &= \mathrm{Tr}\left(\begin{pmatrix} e^{i\nu} & 0 \\ 0 & e^{-i\nu} \end{pmatrix} \int_0^a Y^{-1} JW Y\right). \end{aligned} \tag{1.6.12}$$

Specifically for Hill's equation and the periodic Dirac system, this gives the following result.

Theorem 1.6.2. (a) *Let S be the stability of Hill's equation (1.5.2). Let $\lambda \in S$ and let y be a corresponding complex Floquet solution of (1.5.2). Let $e^{i\nu(\lambda)}$ be the Floquet multiplier, and let*

$$Y = \begin{pmatrix} y & \overline{y} \\ py' & p\overline{y}' \end{pmatrix}$$

be the fundamental matrix formed from y and its complex conjugate. Then

$$D'(\lambda) = \frac{2i \sin \nu(\lambda)}{\det Y(0)} \int_0^a |y|^2 w. \tag{1.6.13}$$

(b) *Let S be the stability set of the Dirac system (1.5.4) with periodic coefficients. Let $\lambda \in S$ and let u be a corresponding complex Floquet solution of (1.5.4). Let $e^{i\nu(\lambda)}$ be the Floquet multiplier, and let*

$$Y = \begin{pmatrix} u_1 & \bar{u}_1 \\ u_2 & \bar{u}_2 \end{pmatrix}$$

be the fundamental matrix formed from u and its complex conjugate. Then

$$D'(\lambda) = \frac{2i \sin \nu(\lambda)}{\det Y(0)} \int_0^a |u|^2. \tag{1.6.14}$$

Proof. (a) For the Sturm-Liouville system,

$$Y^{-1} JW Y = \frac{1}{\det Y} \begin{pmatrix} \bar{y} & p\bar{y}' \\ -y & -py' \end{pmatrix} \begin{pmatrix} w & 0 \\ 0 & 0 \end{pmatrix} \begin{pmatrix} y & \bar{y} \\ py' & p\bar{y}' \end{pmatrix} = \frac{w}{\det Y} \begin{pmatrix} |y|^2 & \bar{y}^2 \\ -y^2 & -|y|^2 \end{pmatrix}.$$

Now (1.6.13) follows from (1.6.12); note that $\det Y$ is a constant Wronskian. It is also purely imaginary, so the factor of the integral in (1.6.13) is real.

(b) For the Dirac system, we have analogously

$$Y^{-1} JW Y = \frac{1}{\det Y} \begin{pmatrix} |u|^2 & \bar{u}^T \bar{u} \\ -u^T u & -|u|^2 \end{pmatrix},$$

writing $|u|^2$ for $u^T \bar{u} = u_1 \bar{u}_1 + u_2 \bar{u}_2$ as usual. □

1.7 The Mathieu equation

From the general properties of the Hill discriminant observed in the preceding section, it is apparent that stability intervals always have a positive length, since the discriminant changes continuously between -2 and 2, either growing or decreasing, in these intervals. However, the instability intervals are bordered by end-points where the discriminant takes the same value, either -2 or 2, and it may happen that the instability interval vanishes by a coalescence of its two end-points in a single point of coexistence.

Nevertheless, there is a remarkable example of a Hill equation for which this never happens, the Mathieu equation

$$y''(x) + (\lambda - 2c \cos 2x) y(x) = 0 \qquad (x \in \mathbb{R}), \tag{1.7.1}$$

with a non-zero real constant c.

Theorem 1.7.1. *The Mathieu equation (1.7.1) has no points of coexistence, for any $c \in \mathbb{R} \setminus \{0\}$.*

Proof. As in (1.5.5), let ϕ_1 and ϕ_2 be the two solutions forming the canonical fundamental system of (1.7.1) and assume on the contrary that both are π-periodic (the case of π-semi-periodicity can be treated in an analogous way). Then the functions $\phi_1(-x)$ and $-\phi_2(-x)$ are solutions of the same initial-value problems as $\phi_1(x)$ and $\phi_2(x)$ respectively, and thus ϕ_1 is an even function, ϕ_2 an odd function. Therefore they can be expanded in Fourier series

$$\phi_1(x) = \frac{a_0}{2} + \sum_{r=1}^{\infty} a_r \cos 2rx,$$

$$\phi_2(x) = \sum_{r=1}^{\infty} b_r \sin 2rx.$$

Term-by-term differentiation and substitution into (1.7.1) gives

$$\left(c\cos 2x - \frac{\lambda}{2}\right) a_0 + \sum_{r=1}^{\infty} \left((4r^2 - \lambda)\cos 2rx + c(\cos 2(r+1)x + \cos 2(r-1)x)\right) a_r = 0$$

and

$$\sum_{r=1}^{\infty} \left((4r^2 - \lambda)\sin 2rx + c\left(\sin 2(r+1)x + \sin 2(r-1)x\right)\right) b_r = 0,$$

showing that

$$c\, a_1 - \frac{\lambda}{2} a_0 = 0 \qquad (1.7.2)$$

and, for $r \in \mathbb{N}$,

$$(4r^2 - \lambda)\, a_r + c\,(a_{r-1} + a_{r+1}) = 0, \qquad (1.7.3)$$
$$(4r^2 - \lambda)\, b_r + c\,(b_{r-1} + b_{r+1}) = 0, \qquad (1.7.4)$$

where b_0 is defined to be zero. Eliminating the $(4r^2 - \lambda)$ term from these two equations yields the identity

$$a_r b_{r+1} - a_{r+1} b_r = a_{r-1} b_r - a_r b_{r-1} \qquad (r \in \mathbb{N}),$$

and so the sequence $(a_{r-1}b_r - a_r b_{r-1})$ is constant. Due to the convergence of the Fourier series, this sequence tends to 0 as $r \to \infty$, and so it must be the null sequence. In particular, its first term vanishes: $a_0 b_1 = 0$. If however $a_0 = 0$, then by (1.7.2) $a_1 = 0$ and, by the recurrence formula (1.7.3), all coefficients a_r vanish, which is impossible since y_1 is non-trivial. Similarly, if $b_1 = 0$, then $b_2 = 0$ and (1.7.4) shows that all coefficients b_r vanish, contradicting the fact that y_2 is non-trivial. □

We shall return to the Mathieu equation in section 3.7, deriving asymptotics for the length of its instability intervals.

As we shall see in section 3.12, the total absence of coexistence is the rule rather than the exception, both for Hill's equation and for periodic Dirac systems. Nonetheless, it is not easy to find explicit examples, and the Mathieu equation even has this property for all non-zero values of the constant c. If $c = 0$, then all instability intervals except $(-\infty, 0)$ vanish.

We conclude this section by mentioning a second differential equation to which the above Fourier method also applies. The equation is

$$(c - \cos 2x)y''(x) + \lambda y(x) = 0,$$

where $|c| > 1$ to ensure that the leading coefficient is nowhere zero, and it can be shown that there is no point λ of coexistence for π-semi-periodic solutions.

1.8 Periodic, semi-periodic and twisted boundary-value problems

In addition to giving rise to a classification of the global qualitative behaviour of solutions of the periodic equation, as seen in section 1.4, Hill's discriminant also provides information about certain boundary-value problems, firstly on a single period interval, and secondly on an interval comprising several complete periods. For the first type of boundary-value problem, the system (1.5.1) is considered on the period interval $[0, a]$, together with a boundary condition of the form

$$u(a) = \omega u(0) \tag{1.8.1}$$

where $|\omega| = 1$. For the *periodic* boundary-value problem, $\omega = 1$ and, for the *semi-periodic* boundary-value problem, $\omega = -1$. The choice of ω on the upper half C_+ of the unit circle (excluding the points ± 1) gives rise to what we call the *ω-twisted periodic boundary-value problem*. This name is motivated by the observation that here both components of u at a are the values at 0 rotated through a fixed angle in the complex plane; the limiting cases where this angle vanishes or becomes a straight angle correspond to the periodic or semi-periodic condition, respectively. We note that ω on the lower half of the unit circle does not produce a new boundary-value problem because we have only to take the complex conjugate of (1.8.1) and then $\overline{\omega} \in C_+$. Specifically for the Sturm-Liouville equation (1.5.2), the boundary condition (1.8.1) gives

$$y(a) = \omega y(0), \qquad (py')(a) = \omega (py')(0). \tag{1.8.2}$$

A number $\lambda \in \mathbb{C}$ is called an *eigenvalue* of a boundary-value problem if there is a non-trivial solution of (1.5.1) satisfying the relevant boundary condition; any such solution is called an *eigenfunction* of this boundary-value problem. An eigenfunction u is said to be *normalised* if $\int_0^a u^T W \overline{u} = 1$, and this can be arranged on multiplying u by a suitable constant. The two propositions which follow give the basic properties of the eigenvalues and eigenfunctions.

Proposition 1.8.1. *For any $\omega \in C_+ \cup \{-1, 1\}$, all eigenvalues of the boundary-value problem (1.5.1), (1.8.1) are real.*

Proof. Let $\lambda \in \mathbb{C}$ be an eigenvalue of the boundary-value problem and u a corresponding eigenfunction. Then, by (1.5.1) and an integration by parts,

$$\lambda \int_0^a (Wu)^T \overline{u} = \int_0^a (u_1' \overline{u_2} - u_2' \overline{u_1} - (Bu)^T \overline{u})$$

$$= (u_1 \overline{u_2} - u_2 \overline{u_1})\Big|_0^a + \int_0^a (u_2 \overline{u_1'} - u_1 \overline{u_2'} - u^T \overline{Bu}) = \overline{\lambda} \int_0^a (Wu)^T \overline{u};$$

the boundary terms cancel, since $|\omega| = 1$ and then $u_j(a)\overline{u_k(a)} = \omega u_j(0)\overline{\omega} \overline{u_k(0)} = u_j(0)\overline{u_k(0)}$. Since u is non-trivial, it follows that $\lambda = \overline{\lambda}$. □

Clearly, any eigenfunction of the periodic or semi-periodic boundary-value problem on $[0, a]$ extends to a non-trivial a-periodic or a-semi-periodic solution of the periodic system on the real line, and conversely any such solution, when restricted to the period interval, yields an eigenfunction. Therefore the eigenvalues of the periodic boundary-value problem are exactly those real values of the spectral parameter λ for which $D(\lambda) = 2$. In the case of coexistence, there are two linearly independent eigenfunctions, and so the eigenvalue has multiplicity 2; in the case of conditional stability the eigenvalue is simple.

Similarly, the eigenvalues of the semi-periodic boundary-value problem are the real values of the spectral parameter λ for which $D(\lambda) = -2$, and the eigenvalue is double if and only if there is coexistence of semi-periodic solutions.

Proposition 1.8.2. *Let λ and μ be distinct eigenvalues of the boundary-value problem (1.5.1), (1.8.1), where $\omega \in C_+ \cup \{-1, 1\}$. Then the corresponding eigenfunctions u and v satisfy the orthogonality relation*

$$\int_0^a u^T W \overline{v} = 0. \tag{1.8.3}$$

Proof. As in the proof of Proposition 1.8.1,

$$\lambda \int_0^a (Wu)^T \overline{v} = \mu \int_0^a (Wu)^T \overline{v},$$

and (1.8.3) follows since $\lambda \neq \mu$ and W is symmetric. □

In the case of coexistence where u and v correspond to the same eigenvalue, we can still arrange to have orthogonality by choosing the two eigenfunctions to be u and $v - cu$, where

$$\int_0^a u^T W(\overline{v} - \overline{cu}) = 0,$$

that is, $\overline{c} = u^T W \overline{v} / u^T W \overline{u}$. In the sequel we always take it that (1.8.3) holds for two linearly independent eigenfunctions.

1.8. Periodic, semi-periodic and twisted boundary-value problems

When (1.5.1) is the Sturm-Liouville or Dirac equation (1.5.2) or (1.5.4), (1.8.3) becomes

$$\int_0^a y\bar{z}w = 0, \quad \int_0^a (u_1\bar{v}_1 + u_2\bar{v}_2) = 0. \tag{1.8.4}$$

In the case of the ω-twisted problem, the eigenvalues are all simple since (1.8.1) implies that the Wronskian of two eigenfunctions for the same eigenvalue is zero. Further, any eigenfunction u must have the form $u = \Phi c$ with a constant non-zero c, and then (1.8.1) gives $\Phi(a)c = \omega c$. Thus ω is an eigenvalue of the monodromy matrix. The other eigenvalue is $\bar{\omega}$ and then (1.4.6) gives $D(\lambda) = 2\operatorname{Re}\omega$. Thus the eigenvalues of this boundary-value problem arise as the pre-images of $2\operatorname{Re}\omega$ under the function D, and hence by Theorem 1.6.1 (a) form a discrete subset of $\mathcal{S}$, with exactly one point in each stability interval. Conversely, it follows from (1.4.8) that

$$\mathcal{S} = \bigcup_{\omega \in C_+} \{\lambda \in \mathbb{R} \mid \lambda \text{ is an } \omega\text{-twisted eigenvalue}\}. \tag{1.8.5}$$

For the second type of boundary-value problem, there is the following connexion between ω-twisted problems on $[0, a]$ and on the larger interval $[0, ka]$, where $k \in \mathbb{N}$.

Theorem 1.8.3. *Let $k \in \mathbb{N}$, $\chi \in C_+ \cup \{-1, 1\}$. Then $\lambda \in \mathbb{R}$ is an eigenvalue of the boundary-value problem for (1.5.1) on $[0, ka]$ with boundary condition $u(ka) = \chi u(0)$ if and only if it is an eigenvalue of the boundary-value problem (1.5.1), (1.8.1) on $[0, a]$, where $\omega \in C_+ \cup \{-1, 1\}$ is a k-th root of χ.*

Proof. If $|D(\lambda)| = 2$, then λ is a periodic or semi-periodic eigenvalue on $[0, a]$, and both sides of the statement hold with $\omega \in \{-1, 1\}$, $\chi = \omega^k$.

Otherwise the two Floquet multipliers are distinct, and there are two linearly independent Floquet solutions (1.4.8). The general solution takes the form

$$u(x) = c_1 e^{\mu x/a} g_1(x) + c_2 e^{-\mu x/a} g_2(x) \quad (x \in \mathbb{R}),$$

where g_1, g_2 are a-periodic and $g_1(0), g_2(0) \in \mathbb{C}^2$ are linearly independent; $\mu \in \mathbb{C}$ is a Floquet exponent and $c_1, c_2 \in \mathbb{C}$ are arbitrary constants. If λ is an eigenvalue for the boundary-value problem on $[0, ka]$, then c_1, c_2 can be chosen such that they are not both zero and

$$c_1 e^{k\mu} g_1(0) + c_2 e^{-k\mu} g_2(0) = u(ka) = \chi u(0) = c_1 \chi g_1(0) + c_2 \chi g_2(0),$$

which by linear independence implies that either $\chi = e^{k\mu}$ or $\chi = e^{-k\mu}$. Hence one of the Floquet multipliers is a k-th root of χ and, calling it ω, we see that the corresponding Floquet solution yields an eigenfunction of the boundary-value problem on $[0, a]$ with (1.8.1).

Conversely, the extension of an eigenfunction of the boundary-value problem on $[0, a]$ satisfies

$$u(x + na) = \omega^n u(x) \quad (x \in [0, a], n \in \mathbb{Z})$$

and thus gives an eigenfunction for the boundary-value problem on $[0, ka]$ for the same eigenvalue λ. □

In particular, Theorem 1.8.3 shows that, if there is a $2a$-periodic solution, that is, $\chi = 1$ and $k = 2$, then $\omega = \pm 1$ and the solution is either a-periodic or a-semi-periodic. Moreover, for $k \geq 3$, the existence of a ka-periodic solution always entails the coexistence of ka-periodic solutions because then there are at least two values of ω which are complex conjugates, providing therefore the coexistence.

Finally, we note a connection with the stability set $\mathcal{S}$. Taking $\omega = 1$ in Theorem 1.8.3 and using the fact that the k-th roots of unity are the elements of $e^{(2\pi i/k)\mathbb{Z}}$, we see that the periodic eigenvalues on $[0, ka]$ are those real values of λ such that $D(\lambda) = 2\cos\dfrac{2n\pi}{k}$ for some integer n; in fact, due to the periodicity of the cosine function, one can always choose $n \in \{1, \ldots, k\}$. As the rationals $\dfrac{n}{k}$ ($n \in \{1, \ldots, k\}$, $k \in \mathbb{N}$) are dense in $[0, 1]$ and D is strictly monotonic in stability intervals by Theorem 1.6.1 (a), it follows that the union Σ of the sets of all eigenvalues of the periodic boundary-value problems on $[0, ka]$ for all $k \in \mathbb{N}$ is a dense subset of the closure of $\mathcal{S}$. Thus $\overline{\Sigma} = \overline{\mathcal{S}}$.

1.9 Appendix: Rofe-Beketov's formula

In this section, which is in the nature of an appendix to the main text, we give further details on Rofe-Beketov's formula, Proposition 1.2.3, which we have used in section 1.4. In particular, we shall obtain this formula as a special case of Theorem 1.9.1 below.

Assume we are given a non-trivial solution u of the system

$$u' = Au \tag{1.9.1}$$

on an interval $X \subset \mathbb{R}$, where $A : X \to \mathbb{C}^{2 \times 2}$, and we wish to derive a formula for a second solution v, expressed in terms of u.

The best-known example of such a formula is d'Alembert's formula for the Sturm-Liouville differential equation

$$-(py')' + qy = 0, \tag{1.9.2}$$

which is equivalent to (1.9.1) with

$$u = \begin{pmatrix} y \\ py' \end{pmatrix}, \quad A = \begin{pmatrix} 0 & \frac{1}{p} \\ q & 0 \end{pmatrix}.$$

Then d'Alembert's formula for a second solution z of (1.9.2) is

$$z(x) := y(x) \int_{x_0}^{x} \frac{1}{py^2}, \tag{1.9.3}$$

1.9. Appendix: Rofe-Beketov's formula

where $x_0 \in X$ is fixed. The Wronskian determinant

$$\begin{vmatrix} y & z \\ py' & pz' \end{vmatrix} = 1,$$

ensuring linear independence of y and z.

The formula (1.9.3) is simple and memorable, but has the flaw that it represents the second solution only on intervals where y is nowhere zero. Indeed, any zero of the non-trivial solution y is simple and thus gives rise to a non-integrable singularity in the d'Alembert integral.

Nevertheless, there is an alternative and less familiar formula with which it is always possible to complement a real-valued, non-trivial solution y of (1.9.2) with a linearly independent solution z, even if y has zeros. This formula, due to Rofe-Beketov [154] (see also [161]) is

$$z(x) := -\frac{(py')(x)}{y^2(x) + (py')^2(x)} + y(x) \int_{x_0}^{x} \frac{(q + \frac{1}{p})(y^2 - (py')^2)}{(y^2 + (py')^2)^2}, \tag{1.9.4}$$

valid throughout X, and again $W(y, z) = 1$.

It is simple to verify that z is a solution of (1.9.2) if y is. However (1.9.4) can be generalised to the system (1.9.1), and this general formula is the subject of the following theorem, due again to Rofe-Beketov [155] (see also [162, Lemma 1] and [39]).

Using the matrix $J = \begin{pmatrix} 0 & 1 \\ -1 & 0 \end{pmatrix}$, we observe that

$$JAJ + A = A^T + A - (\operatorname{Tr} A)I, \tag{1.9.5}$$

Also, $J^2 = -I$ and, for any 2-vector u,

$$u^T J u = 0. \tag{1.9.6}$$

Theorem 1.9.1. *Let u be a non-trivial real-valued solution of (1.9.1). Then a linearly independent solution v of (1.9.1) is defined by*

$$v = fJu + gu, \tag{1.9.7}$$

where the scalar functions f and g are

$$f(x) = -|u(x)|^{-2} \exp\left(\int_{x_0}^{x} \operatorname{Tr} A\right), \tag{1.9.8}$$

$$g(x) = \int_{x_0}^{x} |u(t)|^{-4} u(t)^T \left(JA(t) - A(t)J\right) u(t) \exp\left(\int_{x_0}^{t} \operatorname{Tr} A\right) dt, \tag{1.9.9}$$

and the point x_0 is fixed. The Wronskian $W(u, v)(x_0) = 1$.

Proof. On substituting (1.9.7) into (1.9.1), we find that f and g must satisfy

$$f'Ju + g'u = f(AJ - JA)u. \qquad (1.9.10)$$

Hence, by (1.9.6), $g'u^T u = fu^T(AJ - JA)u$, giving

$$g' = f|u|^{-2} u^T(AJ - JA)u. \qquad (1.9.11)$$

This determines g once f is known. Next, since $J^2 = -I$, (1.9.10) gives

$$-f'u^T u = fu^T(JAJ + A)u.$$

Hence, by (1.9.5)

$$\begin{aligned} f'/f &= -|u|^{-2} u^T(A^T + A - (\operatorname{Tr} A)I)u \\ &= -2|u|'/|u| + \operatorname{Tr} A \end{aligned}$$

since u satisfies (1.9.1). Integrating, we obtain

$$f = (\text{const.})|u|^{-2} \exp\left(\int_{x_0}^{x} \operatorname{Tr} A\right)$$

as in (1.9.8), where we choose the constant to give the Wronskian the desired value, and then (1.9.9) follows from (1.9.11). □

In the particular case where $\operatorname{Tr} A = 0$ (1.9.7) simplifies to

$$v(x) = -|u(x)|^{-2} Ju(x) + u(x) \int_{x_0}^{x} |u|^{-4} u^T(JA - AJ)u, \qquad (1.9.12)$$

thus giving Proposition 1.2.3.

1.10 Chapter notes

§§1.3–1.4 The original paper by Floquet is [59] where differential equations of order n are considered. Standard accounts of Floquet theory for differential equations or, more generally, systems are in Coddington and Levinson [28, pp.78-81], Eastham [48, Chapter 1], Ince [97, pp.381-3], Magnus and Winkler [127, Chapter 1], Stoker [174, Chapter 6], Yakubovich and Starzhinskii [203, Chapter 2, §2]. See also Burnat [27]. For a periodic fourth-order equation, see [10]; for a periodic second-order 2×2 system, see [9].

§1.3 In (1.3.10), $G(x)$ and B are not necessarily real even when $\Psi(x)$ is real. There is however a modification when $\Psi(x)$ is real which states that there are then real matrices $G(x)$, B and E such that (1.3.10) holds with $G(x + a) = G(x)E$, $BE = EB$ and $E^2 = I$. This result is due to Jakubovič [101]; see also Yakubovich

1.10. Chapter notes

and Starzhinskii [203, chapter 1, §2.7]. In particular, G has period $2a$; see [28, p. 81]. In the case $n=2$, there is a further refinement in which G is either a-periodic or a-semi-periodic [203, chapter 8, §1.2].

Theorem 1.3.3 is due to Epstein [55]. Muldowney [138] gives an extension in which there is a more general condition on $F \circ h$.

§1.4 Shi et al. [165, 166, 178] consider the system (1.4.1) with $A = \begin{pmatrix} 0 & 1 \\ -q^2 & 0 \end{pmatrix}$ as a perturbation of the one with (1.3.12). The variation of constants formula and an iterative procedure then provide an algorithm for the computation of D for the Hill equation $y'' + q^2 y = 0$.

In Example 1 the appellation Meissner equation arises from the paper [131] by Meissner (see also the notes on §2.7). A full discussion of the stability of (1.4.10) is given by Almira and Torres [3, Theorem 2].

The differential equation constructed in Example 2 has other applications and is due to Wintner [199, p.394]. Further examples of this general type are constructed by Jakupov [102], where conditions on the constants q_i are given for

$$y''(x) + (q_0 + q_1 \cos x + q_2 \sin x + q_3 \cos 2x) y(x) = 0$$

to have a solution of the form $t(x) \exp(c \cos x)$ with $t(x)$ a trigonometric polynomial. Specific solutions of this type appear in section 2.8 below, Example 2.

§1.5 Djakov and Mityagin [34], [36] introduce the Dirac system in the form

$$i\sigma_3 y' + \begin{pmatrix} 0 & p \\ \bar{p} & 0 \end{pmatrix} y = \lambda y.$$

The transformation

$$y = \frac{1}{2} \begin{pmatrix} -i & 1 \\ 1 & -i \end{pmatrix} u$$

takes this system into the form (1.5.4) with $p_1 = -\operatorname{Im} p$, $p_2 = \operatorname{Re} p$ and $q = 0$, thus giving rise to only a particular case of (1.5.4).

§1.6 For the Sturm-Liouville case, we refer to the notes on section 2.4 below for long-standing accounts of $D(\lambda)$ and the related Theorem 2.4.2. Accounts which include Theorem 1.6.1 for the general system (1.5.1) are in Harris [77] and Yakubovich and Starzhinskii [203, Chapter 7]; see also Starzhinskii [173] and Krein [119].

When $p = w = 1$, an alternative proof of Theorem 1.6.1 for the Sturm-Liouville case is in Hochstadt [87], where λ is taken as a complex parameter and complex variable theory is used.

Khmelnytskaya and Rosu [108] derive a power series representation of $D(\lambda)$ in terms of the first (non-vanishing) periodic eigenfunction, together with supporting numerical computations.

§1.7 Ince [95], [97, §7.4.1]. A number of alternative proofs of Theorem 1.7.1 are in Marković [129], Hille [84], Bremekamp [20] and Bouwkamp [19]. See also Arscott [5, §2.4].

A detailed treatment of the coexistence problem for a variety of equations, including those of Mathieu and Lamé, is given in Magnus and Winkler [127, Chapter 7].

An early conjecture that Theorem 1.7.1 holds for general non-zero even potentials [98], [130] was too optimistic, as pointed out by Borg [17], and a simple example is the Meissner equation in section 2.8 below, Example 1. Another example is provided by the addition of a $\cos 2x$ term: see [110] and the following note on [37].

Djakov and Mityagin [37] developed the method of this section to the two-term potential $q(x) = 2c\cos 2x + b\cos 4x$, and their detailed analysis gives the following results.

A If $b > 0$, the periodic and semi-periodic eigenvalues are all simple (i.e., no coexistence).

B If $b < 0$, define $t = \sqrt{-c^2/2b}$. Then all periodic eigenvalues are simple if t is not an odd integer, and all semi-periodic eigenvalues are simple if t is not an even integer.

C Let b and t be as in **B**. If t is an odd integer, the first t periodic eigenvalues are simple and the others are double. If t is an even integer, the first t semi-periodic eigenvalues are simple and the others are double.

Djakov and Mityagin [36] consider analogues of the Mathieu equation for the Dirac system (1.5.4) in which either

(i) $p_1 = q = 0$, $p_2 = c\cos 2x$ or

(ii) $p_1 = -c\sin 2x$, $p_2 = c(1 + \cos 2x)$, $q = 0$

(see also the note on section 1.5). For (i) the periodic eigenvalues are double and the semi-periodic eigenvalues are simple. For (ii) the periodic and semi-periodic eigenvalues are all simple.

Grigis [70] considers the trigonometric polynomial

$$q(x) = k\cos 2Nx + \sum_{1}^{M}(a_r \cos 2rx + b_r \sin 2rx),$$

where $k > 0$ and $M < N$ with M and N relatively prime. A condition on a_M and b_M is given which implies no coexistence for λ large enough.

§1.8 For the Sturm-Liouville equation, the characterisation (1.8.5) of S is in [48, Theorem 2.4.2]. A similar characterisation holds for complex-valued coefficients and for higher-order equations; see Tkachenko [186] and [187].

1.10. Chapter notes

Theorem 1.8.3. for general χ appears to be new here.

The observation that Σ is a dense subset of S is due to Titchmarsh [183, §21.10]; see also [48, Theorem 2.4.1]. Similar results for Dirac systems are in Harris [77, Theorems 2.8 and 3.8].

There are analogues of the two boundary-value problems in this section for the periodic Schrödinger equation in two or more dimensions with corresponding characterisations of the stability set; see [44] and [48, §§6.2 and 6.5].

Chapter 2

Oscillations

2.1 Introduction

The Floquet theory of Chapter 1 gives an overview of the global growth properties of the solutions of periodic systems. For the purposes of spectral analysis of formally symmetric systems, the oscillations or rotations of the real-valued solutions are also of similar importance. The tool for studying oscillations in Sturm-Liouville and Dirac systems is the Prüfer transform, which is introduced in section 2.2 and then used to analyse the boundary-value problems with separated boundary conditions on the period interval in section 2.3. When conjoined with the results on the periodic and semi-periodic boundary-value problems, this leads to the observation that the oscillations of the solutions of periodic systems have a linear growth asymptotic. The growth rate is a continuous, monotone increasing function of the real spectral parameter. It is known as the rotation number and is connected, by the physical interpretation of the equations, to the quasimomentum and the integrated density of states. In the special case of Hill's equation, the oscillation properties can be equivalently studied by counting zeros of solutions.

We shall continue to use the notation of the general 2×2 system (1.5.1), assuming the general hypotheses on the coefficient matrices B and W. However, from section 2.3 onwards, we shall often make the more specific assumption that this system is either a Sturm-Liouville or a Dirac system. The underlying reason for this is that the general hypotheses on the matrix-valued function W allow it to be singular, as in the case of the Sturm-Liouville system. This leads to specific effects, such as the results in section 2.5, and also means that some proofs, e.g. that of Theorem 2.3.4, employ different techniques for the Sturm-Liouville system and for the Dirac system, and will not extend in a straightforward way to the fully general system (1.5.1).

2.2 The Prüfer transform

Consider now the 2×2 linear periodic system (1.5.1) under the general hypotheses made on the functions B and W in section 1.5, and with spectral parameter $\lambda \in \mathbb{R}$. This makes A an $\mathbb{R}^{2\times 2}$-valued function, and it is therefore sufficient to study the $\mathbb{R}^2$-valued solutions of the system; indeed, a $\mathbb{C}^2$-valued solution can be understood as a complex linear combination of two $\mathbb{R}^2$-valued solutions.

By the uniqueness theorem for solutions of initial-value problems, no non-trivial solution of (1.5.1) takes the value $\begin{pmatrix} 0 \\ 0 \end{pmatrix}$, and so so every non-trivial solution traces out a trajectory in the punctured plane $\mathbb{R}^2 \setminus \{0\}$. Therefore, for any solution $u : X \to \mathbb{R}^2 \setminus \{0\}$, there are locally absolutely continuous functions $\rho : X \to (0, \infty)$ and $\theta : X \to \mathbb{R}$ such that

$$u = \rho \begin{pmatrix} \sin\theta \\ \cos\theta \end{pmatrix}. \qquad (2.2.1)$$

Clearly $\rho = \sqrt{u_1^2 + u_2^2}$ is uniquely determined, and θ is unique up to an additive constant integer multiple of 2π; ρ is called the *Prüfer radius*, and θ the *Prüfer angle*, of the solution u.

Differentiating (2.2.1) and using (1.5.1), we find that

$$\rho' \begin{pmatrix} \sin\theta \\ \cos\theta \end{pmatrix} + \rho\theta' \begin{pmatrix} \cos\theta \\ -\sin\theta \end{pmatrix} = \rho J(B+\lambda W) \begin{pmatrix} \sin\theta \\ \cos\theta \end{pmatrix},$$

and hence we obtain the set of differential equations for the Prüfer variables

$$\theta' = \begin{pmatrix} \sin\theta \\ \cos\theta \end{pmatrix}^T (B+\lambda W) \begin{pmatrix} \sin\theta \\ \cos\theta \end{pmatrix}, \qquad (2.2.2)$$

$$(\log \rho)' = \begin{pmatrix} -\cos\theta \\ \sin\theta \end{pmatrix}^T (B+\lambda W) \begin{pmatrix} \sin\theta \\ \cos\theta \end{pmatrix}. \qquad (2.2.3)$$

The right-hand sides of both equations are quadratic trigonometric polynomials in θ and do not refer to ρ. Consequently, the equation for ρ can be solved by a simple quadrature once θ is known, and the equation for θ is a single non-linear first-order differential equation (known as the *Prüfer equation*) which is essentially equivalent to the linear 2×2 system. We have not so far indicated the dependence of ρ and θ on λ, but we will do so in the next section.

Specifically for the Sturm-Liouville equation (1.5.2), the Prüfer transformation takes the form

$$y = \rho \sin\theta, \qquad py' = \rho \cos\theta, \qquad (2.2.4)$$

and the Prüfer equation becomes

$$\theta' = \frac{1}{p}\cos^2\theta + (\lambda w - q)\sin^2\theta, \qquad (2.2.5)$$

2.2. The Prüfer transform

with the Prüfer radius given by

$$(\log \rho)' = \frac{1}{2}\left(\frac{1}{p} + q - \lambda w\right)\sin 2\theta, \qquad (2.2.6)$$

and hence

$$\rho(x) = \rho(x_0)\exp\left(\int_{x_0}^{x} \frac{1}{2}\left(\frac{1}{p} + q - \lambda w\right)\sin 2\theta\right) \qquad (x, x_0 \in X).$$

For the Dirac system, the Prüfer equation is

$$\theta' = \lambda - q + p_1 \cos 2\theta - p_2 \sin 2\theta, \qquad (2.2.7)$$

and the Prüfer radius satisfies

$$(\log \rho)' = p_1 \sin 2\theta + p_2 \cos 2\theta, \qquad (2.2.8)$$

and therefore can be expressed as

$$\rho(x) = \rho(x_0)\exp\left(\int_{x_0}^{x} (p_1 \sin 2\theta + p_2 \cos 2\theta)\right) \qquad (x, x_0 \in X).$$

Analysis of the Prüfer equation is a powerful tool, because the rotation of the solutions in the punctured phase plane (corresponding to oscillations of solutions around 0 in the case of the Sturm-Liouville equation) turns out to encode much of the spectral properties of the system. However, the Prüfer variables in their original form do not always lead to a differential equation with a direction field which can be interpreted easily. Therefore the applicability and power of the Prüfer method is much enhanced by observing that the phase plane can be subjected to a linear transformation before introducing polar coordinates, with coefficients depending on the spectral parameter λ and on the independent variable of the system. If this linear transformation fixes a direction (typically the u_1 direction) and hence introduces no overall rotation, the resulting generalised Prüfer angle ϕ will have the same asymptotics as the original one, but may satisfy a crucially simplified differential equation.

This equation will generally differ from the particular forms (2.2.5) and (2.2.7), but its right-hand side will always be a homogeneous quadratic trigonometric polynomial of the angle variable, being of the form

$$\phi' = a\cos^2\phi + b\sin\phi\cos\phi + c\sin^2\phi, \qquad (2.2.9)$$

with locally integrable, real-valued functions a, b, c. We call (2.2.9) a *Prüfer-type equation*.

Note that the Prüfer-type equation is equivalent to the generalised Riccati equation; indeed, setting $\zeta = \tan\phi$ gives

$$\zeta' = a + b\zeta + c\zeta^2.$$

However, the solutions of the latter equation have moving singularities, which makes it less convenient to study than equation (2.2.9), whose solutions are locally bounded. Moreover, for the purposes of spectral analysis, the long-range growth of the Prüfer angle or a related variable is usually of greater interest than the local information captured in the Riccati variable ζ.

The transition between Prüfer-type equations is described in the following theorem.

Theorem 2.2.1 (Kepler transformation).
Let $X \subset \mathbb{R}$ be an interval, $\phi, f, g : X \to \mathbb{R}$ locally absolutely continuous functions with $f > 0$. Then, with a suitable choice of branches of arctan,

$$\tilde{\phi} := \arctan(f(\tan \phi + g)) \qquad (2.2.10)$$

is locally absolutely continuous and satisfies

$$\tilde{\phi} \in [(n - \tfrac{1}{2})\pi, (n + \tfrac{1}{2})\pi] \quad \Leftrightarrow \quad \phi \in [(n - \tfrac{1}{2})\pi, (n + \tfrac{1}{2})\pi] \qquad (2.2.11)$$

for all $n \in \mathbb{Z}$. Moreover,

$$\tilde{\phi}' = (\log f)' \sin \tilde{\phi} \cos \tilde{\phi} + fg' \cos^2 \tilde{\phi} + \frac{f \phi'}{\cos^2 \phi + f^2(\sin \phi + g \cos \phi)^2}. \qquad (2.2.12)$$

Proof. We write (2.2.10) as $\tan \tilde{\phi} = f(\tan \phi + g)$. Then (2.2.11) follows immediately, and (2.2.12) follows on differentiation. □

We note that, when ϕ satisfies a Prüfer-type equation (2.2.9), then (2.2.12) takes the same form when (2.2.10) is used in the last term. The Kepler transformation (2.2.10) typically arises when (2.2.1) is modified to

$$u = \rho \begin{pmatrix} f_1 \sin \tilde{\theta} \\ f_2 \cos \tilde{\theta} \end{pmatrix}$$

with f_1 and f_2 (both > 0) at our choice. Then the relation between $\tilde{\theta}$ and θ is $\tan \tilde{\theta} = (f_2/f_1) \tan \theta$.

As an example, with the choice of coefficients $p = w = 1$, the Prüfer equation (2.2.5) for the Sturm-Liouville equation is a relatively complicated function of θ and, even if $q = 0$, does not readily reveal the solutions of the equation, which can be solved explicitly in this simple case. A Kepler transformation with $f = \sqrt{\lambda}$, $g = 0$ gives the generalised Prüfer angle $\tilde{\theta} = \arctan(\sqrt{\lambda} \tan \theta)$ with the differential equation

$$\tilde{\theta}' = \frac{\sqrt{\lambda}(\cos^2 \theta + (\lambda - q) \sin^2 \theta)}{\cos^2 \theta + \lambda \sin^2 \theta} = \sqrt{\lambda} - q \frac{\sin^2 \tilde{\theta}}{\sqrt{\lambda}},$$

which in the case $q = 0$ clearly shows linear growth with slope $\sqrt{\lambda}$ for $\tilde{\theta}$ and hence, up to an error globally bounded by π, for θ as well. For non-zero q, this Prüfer equation is a much better starting point for the derivation of asymptotics than the original Prüfer equation (2.2.5).

In this example, the modifying functions f and g are constant. A more sophisticated use of the Kepler transformation with non-constant coefficient functions appears in the proof of Theorem 2.4.3 below, and this technique will play a central role in Chapter 3.

2.3 The boundary-value problem with separated boundary conditions

The usefulness of the Prüfer transformation for spectral analysis is rooted in the fact that the Prüfer angle depends monotonically on the spectral parameter. This is a consequence of the following basic result on first-order differential inequalitites, due to Chaplygin and Peano.

Theorem 2.3.1. *Let $X \subset \mathbb{R}$, and let $f : X \times \mathbb{R} \to \mathbb{R}$ be locally integrable with respect to the first variable and satisfy the Lipschitz condition*

$$|f(x,y) - f(x,z)| \leq K(x)|y-z|$$

with locally integrable $K : X \to [0, \infty)$. Let $x_0 \in X$ and let ϕ and ψ satisfy the differential inequalities

$$\phi'(x) \leq f(x, \phi(x)), \qquad \psi'(x) \geq f(x, \psi(x))$$

in X, with $\phi(x_0) \leq \psi(x_0)$. Then

(a) *$\phi(x) \leq \psi(x)$ $(x \in X, x \geq x_0)$;*

(b) *if $\phi(x_1) = \psi(x_1)$ for some $x_1 \in X$ and $x_1 > x_0$, then $\phi(x) = \psi(x)$ throughout $[x_0, x_1]$.*

Proof. Let $\delta := \psi - \phi \in AC_{\mathrm{loc}}(X)$. Then $\delta(x_0) \geq 0$ and

$$\delta'(x) = f(x, \psi(x)) - f(x, \phi(x)) \geq -K(x)|\delta(x)|. \qquad (2.3.1)$$

Now assume on the contrary that there is $x' > x_0$ such that $\delta(x') < 0$ and set $x'' := \sup\{x \in [x_0, x') \mid \delta(x) \geq 0\}$. Since δ is continuous, $\delta(x'') = 0$ and $\delta < 0$ on (x'', x'). Therefore, integrating (2.3.1), we have

$$\log \frac{\delta(x')}{\delta(x)} \geq \int_x^{x'} K \qquad (x \in (x'', x')),$$

and hence

$$\delta(x') \geq \lim_{x \to x''} \delta(x) \exp\left(\int_x^{x'} K\right) = 0.$$

This contradiction shows that there is no such x', and part (a) of the theorem follows.

To prove part (b), we again integrate (2.3.1), this time over (x, x_1) with $x \in (x_0, x_1)$. Since now $\delta \geq 0$, this gives

$$\log \frac{\delta(x_1)}{\delta(x)} \geq - \int_x^{x_1} K,$$

and hence $\delta(x) \leq \delta(x_1) \exp\left(\int_x^{x_1} K\right) = 0$. Thus also $\delta \leq 0$ in $[x_0, x_1]$, and part (b) follows. $\square$

Application of this theorem to the differential equation (2.2.2) for the Prüfer angle gives rise to the following statement, which is a generalisation of the classical Sturm Comparison Theorem.

Corollary 2.3.2 (Sturm Comparison). *Let $B_1, B_2, W_1, W_2 : X \to \mathbb{R}^{2 \times 2}$ satisfy the general hypotheses, let $\lambda_1, \lambda_2 \in \mathbb{R}$ and assume that*

$$B_1 + \lambda_1 W_1 - B_2 - \lambda_2 W_2 \geq 0$$

in the sense of positive semidefinite matrices. Let θ_1, θ_2 be Prüfer angles of solutions of

$$u = J(B_1 + \lambda_1 W_1) u, \qquad v = J(B_2 + \lambda_2 W_2) v,$$

respectively, and $\theta_1(x_0) \geq \theta_2(x_0)$ for some $x_0 \in X$. Then $\theta_1(x) \geq \theta_2(x)$ for all $x \in X$, $x \geq x_0$.

This observation proves very useful when comparing differential equations of the type (1.5.1) and in the treatment of perturbations. The most immediate consequence of Theorem 2.3.1 is the monotonic dependence of the Prüfer angle on the spectral parameter. We write (2.2.2) as $\theta'(x) = F(x, \theta, \lambda)$ and consider two different values λ' and λ'' ($\lambda' < \lambda''$) of λ. Since W is positive semidefinite in (2.2.2), $F(x, \theta, \lambda') \leq F(x, \theta, \lambda'')$, and this leads to the following basic property of θ.

Theorem 2.3.3. *Let $\theta(\cdot, \lambda)$ be the Prüfer angle of a real-valued solution of (1.5.1) with $\theta(x_0, \lambda) = \theta_0 \in \mathbb{R}$ and θ_0 is independent of λ. Then, for any fixed $x > x_0$, $\theta(x, \lambda)$ is a strictly increasing continuous function of λ.*

Proof. The continuity follows from (2.2.2) and Theorem 1.2.1 (b). Next, we take $\phi(x) = \theta(x, \lambda')$, $\psi(x) = \theta(x, \lambda'')$ and $f(x, y)$ as $F(x, \theta, \lambda'')$ in Theorem 2.3.1. Then part (a) shows that θ is an increasing function of λ. To show that it is strictly increasing, suppose on the contrary that $\theta(x_1, \lambda') = \theta(x_1, \lambda'')$ for some $x_1 > x_0$. Then, by part (b) of Theorem 2.3.1, $\theta(\cdot, \lambda') = \theta(\cdot, \lambda'')$ on $[x_0, x_1]$. Then again, by (2.2.2),

$$0 = (\lambda' - \lambda'') \begin{pmatrix} \sin \theta \\ \cos \theta \end{pmatrix}^T W \begin{pmatrix} \sin \theta \\ \cos \theta \end{pmatrix}$$

2.3. The boundary-value problem with separated boundary conditions

on $[x_0, x_1]$. This implies that $Wu = 0$ for the non-trivial solution $u = \rho \begin{pmatrix} \sin\theta \\ \cos\theta \end{pmatrix}$, contradicting the general assumption made on W at the beginning of section 1.5. □

The properties of the Prüfer angle provide direct insight into the spectral properties of a class of boundary-value problems for the differential system on a finite interval which, for our purposes, we take to be the period interval $[0, a]$. We also take $\alpha \in [0, \pi)$, $\beta \in (0, \pi]$ and consider the *separated boundary conditions* which for the system (1.5.1) take the general form

$$u_1(0)\cos\alpha - u_2(0)\sin\alpha = 0, \qquad u_1(a)\cos\beta - u_2(a)\sin\beta = 0. \qquad (2.3.2)$$

In the case of the Sturm-Liouville equation (1.5.2), these conditions become

$$y(0)\cos\alpha - (py')(0)\sin\alpha = 0, \qquad y(a)\cos\beta - (py')(a)\sin\beta = 0;$$

here the special case $\alpha = 0$ is called the *Dirichlet boundary condition*, and the case $\alpha = \pi/2$ the *Neumann boundary condition*. At the end-point a, these names are used for $\beta = \pi$ and $\beta = \pi/2$, respectively.

Values of the spectral parameter λ for which there is a non-trivial solution, called an *eigenfunction*, of (1.5.1) satisfying the boundary conditions (2.3.2) are called *eigenvalues* of the boundary-value problem with separated boundary conditions. All such eigenvalues are real; as in the proof of Proposition 1.8.1, the boundary conditions ensure that the boundary terms arising from an integration by parts vanish. Also, the orthogonality property (1.8.4) again holds. The eigenvalues of the boundary-value problems for the Sturm-Liouville equation with $(\alpha, \beta) = (\pi/2, \pi/2)$ and with $(\alpha, \beta) = (0, \pi)$ are briefly referred to as *Neumann eigenvalues* and *Dirichlet eigenvalues*, respectively.

Theorem 2.3.4. (a) *The Sturm-Liouville boundary-value problem on $[0, a]$ with separated boundary conditions has infinitely many real eigenvalues*

$$\lambda_0 < \lambda_1 < \lambda_2 < \ldots$$

with $\lim_{n\to\infty} \lambda_n = \infty$. Any real-valued eigenfunction for the n-th eigenvalue has n zeros in the open interval $(0, a)$.

(b) *The boundary-value problem for the Dirac system on $[0, a]$ with separated boundary conditions has infinitely many eigenvalues*

$$\ldots < \lambda_{-2} < \lambda_{-1} < \lambda_0 < \lambda_1 < \lambda_2 < \ldots$$

with $\lim_{n\to\pm\infty} \lambda_n = \pm\infty$.

In both (a) and (b), the Prüfer angle of any real-valued eigenfunction for λ_n satisfies $\theta(a, \lambda_n) - \theta(0, \lambda_n) = n\pi + \beta - \alpha$.

Proof. (a) For each $\lambda \in \mathbb{R}$, let $\theta(\cdot, \lambda)$ be the Prüfer angle of the solution of the initial-value problem

$$-(py')' + qy = \lambda wy, \qquad \begin{pmatrix} y \\ py' \end{pmatrix}(0, \lambda) = \begin{pmatrix} \sin \alpha \\ \cos \alpha \end{pmatrix},$$

with $\theta(0, \lambda) = \alpha$. By (2.2.4), λ is an eigenvalue if and only if $\theta(a, \lambda) = \beta \pmod{\pi}$. From Theorem 2.3.3, $\theta(a, \cdot)$ is a continuous, strictly increasing function, and so its range is an interval of extent determined by the limits $\lim_{\lambda \to \pm\infty} \theta(a, \lambda)$.

For $\lambda \to -\infty$, we observe first that $\theta(0, \lambda) = \alpha \geq 0$ and that, by (2.2.5), $\theta'(x, \lambda) = \frac{1}{p} > 0$ whenever $\theta(x, \lambda) = 0 \pmod{\pi}$, showing that $\theta(\cdot, \lambda)$ can only increase through the values $n\pi$, $n \in \mathbb{Z}$. Hence $\theta(\cdot, \lambda) \geq 0$, and by monotonicity the limit

$$\theta_{-\infty}(x) := \lim_{\lambda \to -\infty} \theta(x, \lambda) \geq 0$$

exists. We wish to show that the limit is zero for all $x > 0$. For $\lambda \neq 0$, the Prüfer equation (2.2.5) gives

$$\frac{\theta(a, \lambda) - \alpha}{\lambda} = \int_0^a \left(\frac{1}{\lambda} \left(\frac{1}{p} \cos^2 \theta - q \sin^2 \theta \right) + w \sin^2 \theta \right). \qquad (2.3.3)$$

Now $\theta(a, \lambda) < \theta(a, 0)$ when $\lambda < 0$, by Theorem 2.3.3, and so the left-hand side of (2.3.3) tends to zero as $\lambda \to -\infty$. On the right-hand side, as $\lambda \to -\infty$, the integrand tends to $w \sin^2 \theta_{-\infty}$ a.e., and it is bounded by $\frac{1}{p} + |q| + w \in L^1[0, a]$ independently of $\lambda \leq -1$. Hence, by Lebesgue's dominated convergence theorem,

$$0 = \int_0^a w \sin^2 \theta_{-\infty},$$

and so $\theta_{-\infty} = 0 \pmod{\pi}$ a.e. Next, for $\lambda < 0$ and $x \geq t$, (2.2.5) gives

$$\theta(x, \lambda) - \theta(t, \lambda) \leq \int_t^x \left(\frac{1}{p} \cos^2 \theta - q \sin^2 \theta \right).$$

Hence, in the limit $\lambda \to -\infty$,

$$\theta_{-\infty}(x) - \theta_{-\infty}(t) \leq \int_t^x \frac{1}{p}. \qquad (2.3.4)$$

Since $\theta_{-\infty}(a) = \alpha$, (2.3.4) (with $t = a$) shows that $\theta_{-\infty}(x)$ cannot take the value π for x near to a, and hence $\theta_{-\infty}(x) = 0$ for such x. Using (2.3.3) to repeat this argument shows that $\theta_{-\infty}(x)$ can never jump to π, and hence it is zero a.e. as required.

Next, for $\lambda \to \infty$, we show that $\theta(a, \lambda) \to \infty$. We assume on the contrary that $\theta(a, \cdot)$ is bounded; then so is $\theta(x, \cdot) \leq \theta(a, \cdot) + \pi$, and again $\theta_\infty(x) := \lim_{\lambda \to \infty} \theta(x, \lambda)$

2.3. The boundary-value problem with separated boundary conditions

exists for all $x \in [0, a]$. Sending $\lambda \to \infty$ in (2.3.3), we find as before that $\theta_\infty = 0$ (mod π) a.e. On the other hand, for $\lambda > 0$ and $x \geq t$ we have

$$\theta(x, \lambda) - \theta(t, \lambda) \geq \int_t^x \left(\frac{1}{p} \cos^2 \theta - q \sin^2 \theta \right).$$

Hence, in the limit $\lambda \to \infty$,

$$\theta_\infty(x) - \theta_\infty(t) \geq \int_t^x \frac{1}{p} > 0,$$

which shows that θ_∞ is strictly increasing. Since θ_∞ is an integer multiple of π, this contradicts the assumed boundedness of $\theta(a, \cdot)$.

Thus altogether we have shown that the range of $\theta(a, \cdot)$ is $(0, \infty)$. The eigenvalues appear as the pre-images of $\beta + \pi \mathbb{N}_0 \subset (0, \infty)$. The Prüfer angle of the eigenfunction for λ_n will cross 0 (mod π) exactly n times, giving the n zeros of the eigenfunction.

(b) As in (a), let $\theta(\cdot, \lambda)$ be the Prüfer angle of a solution, with $\theta(0, \lambda) = \alpha$. From the Prüfer equation (2.2.7) for the Dirac system, we have

$$|\theta'(\cdot, \lambda) - \lambda| \leq |q| + |p_1| + |p_2|, \tag{2.3.5}$$

and so the total variation of $\theta(x, \lambda) - \lambda x$ ($x \in [0, a]$) is bounded by $C := \|q\|_1 + \|p_1\|_1 + \|p_2\|_1$. In particular,

$$\lambda a - C + \alpha \leq \theta(a, \lambda) \leq \lambda a + C + \alpha.$$

Consequently, the continuous, strictly monotone increasing function $\theta(a, \cdot)$ has range $\mathbb{R}$. The eigenvalues arise as the pre-images of $\beta + \pi \mathbb{Z}$ and can be numbered so that $\theta(a, \lambda_n) = \beta + n\pi$. □

The proof of Theorem 2.3.4 also leads to the following method of counting eigenvalues in an interval.

Theorem 2.3.5 (Relative Oscillation Theorem). *Let $N(\lambda', \lambda'']$ be the number of eigenvalues in the interval $(\lambda', \lambda''] \subset \mathbb{R}$ of a Sturm-Liouville or Dirac boundary-value problem on $[0, a]$ with separated boundary conditions (2.3.2). Let $\theta(\cdot, \lambda)$ be the solution of the associated Prüfer equation with $\theta(0, \lambda) = \alpha$. Then*

$$\frac{1}{\pi}(\theta(a, \lambda'') - \theta(a, \lambda')) - 1 \leq N(\lambda', \lambda''] \leq \frac{1}{\pi}(\theta(a, \lambda'') - \theta(a, \lambda')) + 1.$$

Proof. The number of points λ in the interval $(\lambda', \lambda'']$ where $\theta(b, \lambda) = \beta$ (mod π) is at least $[\frac{1}{\pi}(\theta(a, \lambda'') - \theta(a, \lambda'))]$ and at most $[\frac{1}{\pi}(\theta(a, \lambda'') - \theta(a, \lambda'))] + 1$; here $[x]$ denotes the greatest integer not exceeding x. □

2.4 The rotation number

For the periodic systems, the existence of Floquet solutions allows us to obtain asymptotic information on the growth of Prüfer angles. We first observe that the asymptotics of the Prüfer angle are uniform throughout each instability interval, and are closely related to its counting index. The key idea is to use the eigenvalues of a boundary-value problem with separated boundary conditions on the period interval as tags for the instability intervals. We call the closed intervals of which $\mathbb{R} \setminus \mathcal{S}$ is composed *closed instability intervals*, by a slight abuse of terminology. A closed instability interval either consists of a single coexistence point, or it is the closure of an instability interval.

Theorem 2.4.1. *Let $(\lambda_n)_{n \in \mathcal{J}}$ be the ordered set of eigenvalues for the boundary-value problem on $[0, a]$ for the Sturm-Liouville or Dirac system with Neumann boundary conditions (2.3.2), in which $\alpha = \beta = \pi/2$. Then there is a monotonic enumeration $(\overline{I}_n)_{n \in \mathcal{J}}$ of the closed instability intervals such that $\lambda_n \in \overline{I}_n$ $(n \in \mathcal{J})$.*

For every $\lambda \in \overline{I}_n$, the Prüfer angle of any real-valued solution of the system with spectral parameter λ satisfies

$$\theta(x, \lambda) = \frac{n\pi x}{a} + O_{\mathrm{unif}}(1) \qquad (x \to \infty). \tag{2.4.1}$$

The O_{unif} term is bounded uniformly with respect to $\lambda \in \overline{I}_n$.

As shown in Theorem 2.3.4, $\mathcal{J} = \mathbb{N}_0$ for the Sturm-Liouville system and $\mathcal{J} = \mathbb{Z}$ for the Dirac system.

Proof. By the last statement of Theorem 2.3.4, the Prüfer angle for an eigenfunction for λ_n satisfies

$$\theta_n(a) - \theta_n(0) = n\pi \qquad (n \in \mathcal{J}). \tag{2.4.2}$$

Now the eigenfunction is a multiple of the first column in the canonical fundamental matrix Φ and, in view of the Neumann boundary condition at 0, the monodromy matrix has lower left entry $M_{21}(\lambda_n) = 0$. Hence the product of its diagonal elements is equal to 1. By (1.6.2), it follows that $|D(\lambda_n)| \geq 2$, and we also have $\operatorname{sgn} D(\lambda_n) = (-1)^n$. The alternating sign shows that λ_n and λ_{n+1} lie in distinct closed instability intervals. Therefore the interval $[\lambda_n, \lambda_{n+1}]$ contains at least one stability interval.

Because of the continuity of D, there are points

$$\Lambda_p, \Lambda_s \in [\lambda_n, \lambda_{n+1}] \tag{2.4.3}$$

such that $D(\Lambda_p) = 2$ and $D(\Lambda_s) = -2$. Let Θ_p and Θ_s be Prüfer angles of the (respectively, periodic and semi-periodic) Floquet solutions for Λ_p and Λ_s with initial values $\Theta_p, \Theta_s \in [\frac{\pi}{2}, \frac{5\pi}{2})$.

From (2.4.2) and the periodicity of the system, it follows that

$$\theta_n(x) = \frac{n\pi x}{a} + O(1), \qquad \theta_{n+1}(x) = \frac{(n+1)\pi x}{a} + O(1) \qquad (x \to \infty), \tag{2.4.4}$$

2.4. The rotation number

where the $O(1)$ term represents the variation of the angle in each period interval; for the Sturm-Liouville system it is bounded by π, as the Prüfer angle cannot decrease through $\mathbb{Z}\pi$, for the Dirac system it is bounded uniformly in $[\lambda_n, \lambda_{n+1}]$ because of the integrability of the coefficients, as in (2.3.5).

Since Θ_p and Θ_s are a-periodic modulo π, a similar argument gives

$$\Theta_p(x) = \frac{m_p \pi x}{a} + O(1), \qquad \Theta_s(x) = \frac{m_s \pi x}{a} + O(1) \qquad (x \to \infty) \qquad (2.4.5)$$

for some $m_p \in 2\mathbb{Z}, m_s \in 2\mathbb{Z} + 1$. The $O(1)$ terms are uniformly bounded as above.

Now Theorem 2.3.3 implies, on the basis of (2.4.3) and the initial values for the Prüfer angles, that

$$\Theta_p(x), \Theta_s(x) \in [\theta_n(x), \theta_{n+1}(x) + 2\pi] \qquad (x \geq 0).$$

Comparing the asymptotics (2.4.4) and (2.4.5), this shows that $m_p, m_s \in \{n, n+1\}$. Specifically, if n is even, then $m_p = n$, $m_s = n + 1$; if n is odd, then $m_s = n$, $m_p = n + 1$.

The above reasoning holds true for all $\Lambda_p, \Lambda_s \in [\lambda_n, \lambda_{n+1}]$ such that $D(\Lambda_p) = 2$ and $D(\Lambda_s) = -2$. Hence, by Theorem 2.3.3, all Λ_p are either greater or less than all Λ_s, which will contradict the properties of Hill's discriminant in Theorem 1.6.1 unless there is only one Λ_p and one Λ_s with these properties. Thus we have shown that $[\lambda_n, \lambda_{n+1}]$ contains exactly one stability interval.

At the two end-points of $\overline{I}_n$ we thus have periodic or semi-periodic solutions with Prüfer angles satisfying (2.4.1). By Theorem 2.3.3, this asymptotic holds for all solutions for all $\lambda \in \overline{I}_n$. $\square$

The analysis in the proof of Theorem 2.4.1 reveals the following fundamental properties of the periodic and semi-periodic eigenvalues of Sturm-Liouville and Dirac systems.

Theorem 2.4.2. *The periodic eigenvalues $(\lambda_n)_{n \in \mathcal{J}}$ and the semi-periodic eigenvalues $(\mu_n)_{n \in \mathcal{J}}$ of the Sturm-Liouville ($\mathcal{J} = \mathbb{N}_0$) or Dirac ($\mathcal{J} = \mathbb{Z}$) system on $[0, a]$ form interlacing sequences*

$$(\cdots \leq) \lambda_0 < \mu_0 \leq \mu_1 < \lambda_1 \leq \lambda_2 < \mu_2 \leq \mu_3 < \lambda_3 \leq \lambda_4 < \ldots \qquad (2.4.6)$$

The Prüfer angle of a real-valued eigenfunction for a periodic or semi-periodic eigenvalue λ satisfies

$$\theta(a, \lambda) = \theta(0, \lambda) + \begin{cases} 2(m+1)\pi & \text{if } \lambda \in \{\lambda_{2m+1}, \lambda_{2m+2}\}, \\ (2m+1)\pi & \text{if } \lambda \in \{\mu_{2m}, \mu_{2m+1}\}, \end{cases} \qquad (2.4.7)$$

for any suitable integer m. Moreover,

$$\mathcal{S} = (\cdots \cup) (\lambda_0, \mu_0) \cup (\mu_1, \lambda_1) \cup (\lambda_2, \mu_2) \cup \ldots \qquad (2.4.8)$$

and $\overline{I}_n = [\lambda_{n-1}, \lambda_n]$ for even n, $\overline{I}_n = [\mu_{n-1}, \mu_n]$ for odd n.

Theorem 2.4.1 shows that, for $\lambda \in \bar{I}_n$ and the Prüfer angle of any solution, the limit
$$k(\lambda) := \lim_{x \to \infty} \frac{\theta(x, \lambda)\, a}{x} = n\pi$$
exists. The next theorem will show that this function k extends to the stability intervals. It is a non-decreasing continuous function called the *rotation number*.

Theorem 2.4.3. *Let $\lambda \in S$ and let D be Hill's discriminant of a Sturm-Liouville or Dirac system. Then there is $k(\lambda) \in \mathbb{R}$ such that the Prüfer angle $\theta(x, \lambda)$ of any solution of the system with spectral parameter λ satisfies*
$$\theta(x, \lambda) = \frac{k(\lambda)\, x}{a} + O(1) \qquad (x \to \infty) \tag{2.4.9}$$
and $D(\lambda) = 2\cos k(\lambda)$.

Proof. Since $|D(\lambda)| < 2$, we are in Case 3 of the classification in section 1.4. The Floquet multipliers are $e^{\pm i \nu(\lambda)}$ with $\nu \in \mathbb{R} \setminus \pi \mathbb{Z}$, and the corresponding complex-valued Floquet solutions u and $\bar{u}$ are linearly independent with
$$u(x+a, \lambda) = e^{i\nu(\lambda) x/a} u(x, \lambda). \tag{2.4.10}$$
We will show that $k(\lambda)$ is related to $\nu(\lambda)$.

We begin by considering the components u_1 and u_2 of u. First, they have no zeros because of the linear independence of u and $\bar{u}$. Also, their arguments are nowhere equal mod π, because otherwise we could multiply u by a suitable complex constant c to make both components of cu real at that point, again contradicting the linear independence of u and $\bar{u}$. Hence, either $\arg u_1 < \arg u_2 < \pi + \arg u_1$, or $-\pi + \arg u_1 < \arg u_2 < \arg u_1$. Thus we can write
$$u_1 = R_1 e^{i\phi_1}, \quad u_2 = \pm i R_2 e^{i\phi_2}, \tag{2.4.11}$$
where $R_j (>0)$ and ϕ_j are real-valued absolutely continuous functions with
$$|\phi_1 - \phi_2| < \frac{\pi}{2}. \tag{2.4.12}$$
Without loss of generality we take the plus sign in (2.4.11).

Now let v be any real-valued, non-trivial solution of the differential system with Prüfer angle θ. Thus $v = c(u + \bar{u}) + id(u - \bar{u})$ with real constants c and d, not both zero. By (2.2.1) and (2.4.11),
$$\tan \theta = \frac{R_1}{R_2} \frac{-c \cos \phi_1 + d \sin \phi_1}{c \sin \phi_2 + d \cos \phi_2} = \frac{R_1}{R_2} \frac{\sin(\phi_1 - \gamma)}{\cos(\phi_2 - \gamma)}$$
for some $\gamma \in (\frac{\pi}{2}, \frac{\pi}{2}]$. Hence
$$\tan \theta = \frac{R_1}{R_2} \cos(\phi_1 - \phi_2)\{\tan(\phi_2 - \gamma) + \tan(\phi_1 - \phi_2)\}.$$

2.4. The rotation number

By (2.4.12) $\cos(\phi_1-\phi_2) > 0$ and $\tan(\phi_1-\phi_2)$ is bounded. Then, by Theorem 2.2.1, θ and $\phi_2 - \gamma$ take the values in $(\mathbb{Z} + \frac{1}{2})\pi$ at the same points, and consequently their difference $\theta - \phi_2 + \gamma$ is globally bounded.

From (2.4.10) and (2.4.11), R_1 and R_2 are a-periodic and

$$\phi_2(x+a,\lambda) = \phi_2(x,\lambda) + \nu(\lambda).$$

Thus the function $\phi_2(x,\lambda) - \nu(\lambda)x/a$ is a-periodic and continuous, and therefore globally bounded. By what we have just proved $\theta(x,\lambda) - \nu(x,\lambda)x/a$ has the same property and (2.4.9) follows with $k = \nu$. The relation $D = 2\cos k$ follows from (1.4.6). □

In addition to describing the rate of oscillation of solutions of the periodic equation, the rotation number for a one-dimensional Schrödinger equation or Dirac system has a further physical interpretation. Let u be a Floquet solution for spectral parameter $\lambda \in \mathcal{S}$, so that $u(a) = e^{\mu(\lambda)}u(0)$, and consider $v(x) := e^{-i\kappa x}u(x)$ with a parameter $\kappa \in \mathbb{R}$. Then v will be a-periodic if $\mu(\lambda) = i\kappa a$, which in the light of Theorem 2.4.3 is equivalent to choosing

$$\kappa = \frac{k(\lambda)}{a}.$$

In the case of a one-dimensional Schrödinger equation, the first component v_1 satisfies the modified equation

$$\left(-i\frac{d}{dx} + \kappa\right)^2 y(x) + q(x)\,y(x) = \lambda\, y(x).$$

In the case of the Dirac system, v satisfies

$$\sigma_2\left(-i\frac{d}{dx}+\kappa\right)v(x) + p_1\sigma_3\,v(x) + p_2\sigma_1\,v(x) + q(x)\,v(x) = \lambda\,v(x).$$

In both cases the quantum mechanical momentum operator $-i\frac{d}{dx}$ is replaced with $-i\frac{d}{dx} + \kappa$. Instead of a twisted periodic boundary-value problem on $[0,a]$ for the original periodic equation, one can therefore equivalently consider a periodic boundary-value problem for the modified equation, and this point of view is preferred in physics. Due to this interpretation, κ is called *quasi-momentum*.

The rotation number plays the following further role in quantum mechanics. Consider the boundary-value problem for the a-periodic Sturm-Liouville or Dirac system with separated boundary conditions on an interval $[0,x]$. Let $\theta(\cdot,\lambda)$ be the Prüfer angle for spectral parameter λ such that $\theta(0,\lambda) = \alpha$, where α is the parameter of the boundary condition at 0. By Theorem 2.3.5 and the formulae (2.4.1) and (2.4.9), the asymptotic number of eigenvalues in an interval $(\lambda', \lambda'']$ is given by

$$\frac{1}{\pi}(\theta(\lambda'',x) - \theta(\lambda',x)) \sim \frac{k(\lambda'') - k(\lambda')}{\pi a}x \quad (x \to \infty).$$

Thus the number of eigenvalues (corresponding to physical bound states) grows asymptotically linear with the length of the interval, and it makes sense to interpret

$$\frac{k(\lambda)}{\pi a}$$

as the *integrated density of states*.

Theorem 2.4.3 and Theorem 1.6.2 yield a formula for the derivative of the rotation number in the stability intervals of Hill's equation and the Dirac system. Indeed, the identity $D(\lambda) = 2\cos k(\lambda)$ implies that

$$D'(\lambda) = -2k'(\lambda)\sin k(\lambda) \qquad (2.4.13)$$

and, bearing in mind that $k(\lambda)$ plays the role of $\nu(\lambda)$ in (1.6.13), we find that

$$k'(\lambda) = \frac{1}{i \det Y(0)} \int_0^a |y|^2 w, \qquad (2.4.14)$$

where y is a complex Floquet solution for $\lambda \in \mathcal{S}$ and Y is the fundamental matrix formed from y and $\bar{y}$. (For the Dirac system, read $w = 1$.)

Equation (2.4.13) shows that the derivative of the rotation number diverges at the points of transition from stability to instability. Thus k is continuously differentiable on $\mathcal{S}$ and constant on each component interval of $\mathcal{I}$, but it is not a continuously differentiable function throughout. For Hill's equation, the derivative of the rotation number is bounded below in the stability intervals as follows.

Theorem 2.4.4. *The rotation number k of Hill's equation (1.5.2) satisfies*

$$(k^2)'(\lambda) \geq \left(\int_0^a \sqrt{\frac{w}{p}}\right)^2 \qquad (\lambda \in \mathcal{S}).$$

Proof. Let $\lambda \in \mathcal{S}$ and y a corresponding (complex-valued) Floquet solution of (1.5.2). Then we can write

$$y(x) = |y|(x)\, e^{i\phi_1(x)} \qquad (x \in \mathbb{R}) \qquad (2.4.15)$$

as in (2.4.11) and calculate the rotation number from a Prüfer angle θ of the solution $\operatorname{Re} y$ according to (2.4.9). As shown in the proof of Theorem 2.4.3, $\theta - \phi_1$ is universally bounded, and so

$$k(\lambda) = \lim_{x \to \infty} \frac{\theta(x)\, a}{x} = \lim_{x \to \infty} \frac{\phi_1(x)\, a}{x}. \qquad (2.4.16)$$

Now differentiating (2.4.15) gives

$$y'(x) = (|y|'(x) + i\phi_1'(x)\,|y|(x))\, e^{i\phi_1(x)},$$

2.4. The rotation number

and hence the (constant) Wronskian of the fundamental matrix formed from y and $\bar{y}$ is
$$\det Y = y\, p\bar{y}' - py'\, \bar{y} = -2i\, p\phi_1'\, |y|^2.$$

Writing ϕ_1 as the integral over its derivative, we hence conclude from (2.4.16) that
$$k(\lambda) = \lim_{x\to\infty} \frac{a}{x} \int_0^x \frac{i\, \det Y(0)}{2p\,|y|^2} = i\, \det Y(0) \int_0^a \frac{1}{2p\,|y|^2}.$$

Combining this with (2.4.14), the Wronskian cancels out and we obtain
$$(k^2)'(\lambda) = 2\, k(\lambda)\, k'(\lambda)$$
$$= \int_0^a \frac{1}{|y|^2\, p} \int_0^a |y|^2\, w \geq \left(\int_0^a \sqrt{\frac{w}{p}} \right)^2$$

by the Cauchy-Schwarz inequality (note that the integral on the right always exists, since w and $1/p$ are assumed integrable over $[0,a]$). □

The proof of Theorem 2.4.1 set out from the observation that the eigenvalues of a particular boundary-value problem on $[0, a]$ with separated boundary conditions lie in closed instability intervals. This holds true for all boundary-value problems where the boundary conditions at the two end-points are the same, even if the period interval is shifted by an offset τ to $[\tau, \tau + a]$, as the following result shows. Clearly the eigenvalues of the periodic and semi-periodic boundary-value problems, and consequently the positions of the stability intervals of which they are end-points, remain unchanged under such a shift.

Theorem 2.4.5. (a) *Let $\tau \in [0, a)$. Then the n-th eigenvalue of the boundary-value problem for the Sturm-Liouville or Dirac system on $[\tau, \tau + a]$ with boundary conditions (2.3.2), in which $\alpha = \beta \in (0, \pi)$, lies in the closed instability interval $\bar{I}_n$.*

In the Dirichlet case where $\alpha = 0$ and $\beta = \pi$, the n-th eigenvalue lies in the closed instability interval $\bar{I}_{n+1}$.

(b) *Let $\lambda \in \bar{I}_n$, $n \in \mathcal{J} \setminus \{0\}$, and let $\alpha = \beta \in (0, \pi)$. Then there is $\tau \in [0, a)$ such that λ is the n-th eigenvalue of the boundary-value problem for the Sturm-Liouville or Dirac system on $[\tau, \tau + a]$ with boundary conditions (2.3.2).*

The same holds true in the Dirichlet case $\alpha = 0$ and $\beta = \pi$, except that λ is the $(n-1)$-th eigenvalue.

Proof. (a) Let u be a real-valued eigenfunction, and define another solution v of the system by $v(x) = u(x + a)$. Then, by (2.3.2) referred to $[\tau, \tau + a]$,
$$u_1(\tau) \cos\alpha - u_2(\tau) \sin\alpha = 0$$
and
$$v_1(\tau) \cos\alpha - v_2(\tau) \sin\alpha = 0.$$

Hence the Wronskian of u and v is zero at τ, making u and v linearly dependent. Thus $u(x + a) = cu(x)$ with a real constant c. If now $|c| = 1$, u is either periodic or semi-periodic. If however $|c| \neq 1$, then u is exponentially unbounded at ∞ or $-\infty$, and then the periodic equation is not stable. Either way, the eigenvalue lies in a closed instability interval.

The statement about the eigenvalue numbering follows from the comparison of the growth of the Prüfer angle of u, obtained in Theorem 2.3.4, with the asymptotic (2.4.1).

(b) Since $\lambda \in \bar{I}_n$, there is a real-valued Floquet solution u satisfying (1.3.5). As g is periodic and the exponential factor is positive, the Prüfer angle of u satisfies $\theta(x + a, \lambda) = n\pi + \theta(x, \lambda)$, from (2.4.1). Since $n \neq 0$, we have in particular either $\theta(a, \lambda) \geq \pi + \theta(0, \lambda)$ or $\theta(a, \lambda) \leq -\pi + \theta(0, \lambda)$. The Intermediate Value Theorem then ensures the existence of a point $\tau \in [0, a)$ such that $\theta(\tau, \lambda) = \alpha \pmod{\pi}$ as required. $\square$

The theorem can also be expressed as follows. Let $\Lambda_{\tau,n}$ denote the n-th eigenvalue in the above boundary-value problems on $[\tau, \tau + a]$. Then the range of $\Lambda_{\tau,n}$, as a function of τ, is $\bar{I}_{n+1}$ in the Dirichlet case, and $\bar{I}_n$ otherwise.

2.5 Zeros of solutions of Hill's equation

The Sturm-Liouville equation has the special property that the zeros of real-valued solutions reflect the growth of the corresponding Prüfer angle. Indeed, it is clear from the Prüfer transformation formulae (2.2.4) that these zeros occur exactly at the points where the Prüfer angle vanishes modulo π, and the Prüfer equation (2.2.5) shows that the angle is strictly increasing at such points. Hence the Prüfer angle increases by exactly π between any two adjacent zeros of a solution.

Bearing this in mind, Theorem 2.4.2 provides the following information about the number of zeros of eigenfunctions of the boundary-value problems with periodic or semi-periodic boundary conditions. (Theorem 2.3.4 (a) has the analogous statement for the problem with separated boundary conditions.)

Theorem 2.5.1. (a) *Any eigenfunction for the periodic eigenvalue λ_0 has no zeros in $[0, a]$. Any eigenfunction for the periodic eigenvalues λ_{2m+1} and λ_{2m+2} has $2(m + 1)$ zeros in $[0, a)$ $(m \in \mathbb{N}_0)$.*

(b) *Any eigenfunction for the semi-periodic eigenvalues μ_{2m} and μ_{2m+1} has $2m + 1$ zeros in $[0, a)$ $(m \in \mathbb{N}_0)$.*

For general real-valued solutions of Hill's equation on $\mathbb{R}$ we can draw the following conclusions.

Theorem 2.5.2. *Let λ_0 be the smallest periodic eigenvalue. For $\lambda \in \mathbb{R}$, let y be a non-trivial real-valued solution of (1.5.2). If $\lambda > \lambda_0$, then y has infinitely many zeros. If $\lambda \leq \lambda_0$, then y has at most one zero.*

2.5. Zeros of solutions of Hill's equation

Proof. As $\lambda_0 = \inf \mathcal{S}$, we know that the rotation number $k(\lambda) > 0$ if $\lambda > \lambda_0$, and the first statement follows from the asymptotics (2.4.1), (2.4.9).

If $\lambda < \lambda_0$, assume there are points $x_1 < x_2$ such that $y(x_1) = y(x_2) = 0$. The Prüfer angle θ of y can be normalised so that $\theta(x_1) = 0$; then $\theta(x_2) = n\pi$ with some $n \in \mathbb{N}$. Let θ_0 be the Prüfer angle of a periodic eigenfunction for λ_0; then by Theorem 2.4.2, θ_0 is a-periodic and (without loss of generality) takes values in $(0, \pi)$. By the intermediate value theorem, there is a point $x_0 \in (x_1, x_2)$ such that $\theta(x_0) = \theta_0(x_0)$. But also $\theta(x_2) > \theta_0(x_2)$, contradicting Theorem 2.3.3. □

The statements of Theorem 2.5.2 are sometimes expressed by saying that Hill's equation is *oscillatory* if $\lambda > \lambda_0$, and *disconjugate* if $\lambda < \lambda_0$.

Studying the zeros of the periodic eigenfunctions gives rise to the following upper bounds to the periodic eigenvalues at the lower ends of stability intervals.

Theorem 2.5.3. *The smallest periodic eigenvalue of Hill's equation satisfies*

$$\lambda_0 \leq \left(\int_0^a q \right) \Big/ \left(\int_0^a w \right),$$

with equality if and only if $q = \lambda_0 w$ a.e.

Proof. By Theorem 2.5.1 (a), the periodic eigenfunction y for eigenvalue λ_0 has no zeros and can therefore be taken to be positive. Then, by (1.5.2),

$$(p(\log y)')' = q - \lambda_0 w - \frac{(py')^2}{py^2}. \tag{2.5.1}$$

The left-hand side is the derivative of an a-periodic function, and so integration over a period interval gives

$$\int_0^a q - \lambda_0 \int_0^a w = \int_0^a \frac{(py')^2}{py^2} \geq 0,$$

and hence the stated estimate.

Equality implies that $py' = 0$, and thus $p(\log y)' = 0$ throughout, and the last statement follows in view of (2.5.1). □

Theorem 2.5.4. *Let $m \in \mathbb{N}_0$ and assume that*

$$\int_0^a q(x) e^{2\pi r x i/a}\, dx = 0 \qquad (r \in \{0, \ldots, 2m\}). \tag{2.5.2}$$

Then the periodic eigenvalue λ_{2m} of Hill's equation satisfies

$$\lambda_{2m} \leq \frac{\sup p}{\inf w} (2m\pi/a)^2,$$

with equality if and only if $q = 0$ and p, w are constant a.e.

Proof. Let y be a real-valued eigenfunction for λ_{2m}. By Theorem 2.5.1 (a), y has $2m$ zeros $x_1, \ldots x_{2m} \in [0, a)$. We introduce the function

$$f(x) := \prod_{r=1}^{2m} \sin(\pi(x - x_r)/a)$$

which is a-periodic and has a finite Fourier expansion of the form

$$\sum_{r=-m}^{m} c_r \exp(2\pi r x i / a).$$

Hence

$$\int_0^a f'^2 \leq (2\pi m/a)^2 \int_0^a f^2. \tag{2.5.3}$$

Also, by (2.5.2),

$$\int_0^a q f^2 = 0. \tag{2.5.4}$$

The function y/f (extended by continuity at each x_r) is a-periodic and (without loss of generality) positive. Then, with $g := \log(y/f)$, we can write $y = fe^g$. Multiplication of Hill's equation (1.5.2) by fe^{-g} and integration over the period interval gives

$$\int_0^a (q - \lambda_{2m} w) f^2 = \int_0^a fe^{-g} (py')' = fe^{-g}(py')\Big|_0^a + \int_0^a p(fg')^2 - \int_0^a p f'^2. \tag{2.5.5}$$

The boundary terms cancel by periodicity and, using (2.5.4), we find that

$$\lambda_{2m} \leq \int_0^a p f'^2 \Big/ \int_0^a w f^2,$$

and the stated estimate follows by (2.5.3).

In the case of equality, it follows that p and w must be constant in view of the last integral estimates. Moreover, from (2.5.5), $g' = 0$ and hence g is constant. Also to make (2.5.3) sharp, f must be a linear combination of $e^{\pm 2m\pi x i/a}$ only. This means that $y(x) = \sin(2m\pi x/a + \phi)$ with a constant $\phi \in \mathbb{R}$. Using this information along with $\lambda_{2m} = p(2m\pi/a)^2/w$ in Hill's equation gives $q = 0$ a.e. □

There is also a corresponding result to Theorem 2.5.4 for the semi-periodic eigenvalue μ_{2m+1} in which $2m$ is replaced by $2m + 1$.

2.6 The upper end-points of the stability intervals

The eigenvalues λ_{2m} and μ_{2m+1} in the previous section are the lower end-points of the stability intervals. In this section we consider the upper end-point $\lambda^{(n)}$ of the

2.6. The upper end-points of the stability intervals

n-th stability interval. Thus $\lambda^{(n)}$ is μ_{n-1} or λ_{n-1} according as n is odd or even. As a further application of Theorem 2.5.1, we give a lower bound for $\lambda^{(n)}$. In the following theorem, we write

$$A = \int_0^a p^{-1} \tag{2.6.1}$$

and

$$M = \sup(pw) \quad \text{in } [0, a].$$

Theorem 2.6.1. *Let*

$$\int_0^a q_- \geq -4n^2/A, \tag{2.6.2}$$

where $q_-(x) = \min\{q(x), 0\}$ and n is a positive integer. Then

$$\lambda^{(n)} \geq M^{-1}(n\pi/A)^2 \left(1 + \frac{A}{4n^2}\int_0^a q_-\right). \tag{2.6.3}$$

Equality holds in (2.6.3) only when $q = 0$ and $pw = M$ a.e.

Proof. Let y be the eigenfunction corresponding to $\lambda^{(n)}$, so that y is a-periodic or a-semi-periodic as the case may be. Let $x_0, \ldots, x_n$ be $n+1$ consecutive zeros of y in the order $x_0 < x_1 < \cdots < x_n$. By Theorem 2.5.1, we have

$$x_n = x_0 + a. \tag{2.6.4}$$

In Hill's equation (1.5.2), we write $\lambda = \lambda^{(n)}$ and then multiply by y and integrate over $[x_r, x_{r+1}]$. After an integration by parts, this gives

$$\lambda^{(n)} \int_{x_r}^{x_{r+1}} y^2 w = \int_{x_r}^{x_{r+1}} (py'^2 + qy^2). \tag{2.6.5}$$

Also,

$$2y(x) = \int_{x_r}^x y' - \int_x^{x_{r+1}} y'$$

giving

$$2|y(x)| \leq \int_{x_r}^{x_{r+1}} |y'|$$

for all x in $[0, a]$. Hence, by the Cauchy-Schwarz inequality,

$$y^2(x) \leq \tfrac{1}{4}(x_{r+1} - x_r) \int_{x_r}^{x_{r+1}} y'^2. \tag{2.6.6}$$

Let us for the moment consider the special case when $p(x) = 1$, so that $A = a$ in (2.6.1). Then (2.6.5), (2.6.6) and the inequality $q \geq q_-$ give

$$\lambda^{(n)}/\rho_r \geq (x_{r+1} - x_r)^{-1} + \tfrac{1}{4}\int_{x_r}^{x_{r+1}} q_-, \tag{2.6.7}$$

where
$$\rho_r = (x_{r+1} - x_r) \int_{x_r}^{x_{r+1}} y'^2 \Big/ \int_{x_r}^{x_{r+1}} y^2 w. \qquad (2.6.8)$$

We now sum (2.6.7) for $r = 0, \ldots, n-1$ and use the harmonic mean/arithmetic mean inequality

$$\sum_{r=0}^{n-1} (x_{r+1} - x_r)^{-1} \geq n^2 \left(\sum_{r=0}^{n-1} (x_{r+1} - x_r) \right)^{-1}.$$

Then (2.6.7) gives

$$\lambda^{(n)} \sum_{r=0}^{n-1} \rho_r^{-1} \geq n^2 a^{-1} + \tfrac{1}{4} \int_0^a q_-, \qquad (2.6.9)$$

where we have used (2.6.4) and the periodicity of q_-. By (2.6.3) (with $A = a$), the right-hand side of (2.6.9) is non-negative, and hence $\lambda^{(n)} \geq 0$.

At this point, we require the Wirtinger inequality which states that

$$\int_c^d |f'|^2 \geq \frac{\pi}{(d-c)^2} \int_c^d |f|^2 \qquad (2.6.10)$$

for any f which is $AC[c,d]$ with $f(c) = f(d) = 0$. Applying this inequality in (2.6.8), along with $w \leq M$, we have

$$\rho_r \geq M^{-1} \pi^2 (x_{r+1} - x_r)^{-1}. \qquad (2.6.11)$$

Then (2.6.9) yields (2.6.3) with $A = a$.

To deal with general $p(x)$, we make the change of variable

$$u = \int_0^x p^{-1},$$

so that (1.5.2) becomes

$$-d^2 Y(u)/du^2 + Q(u) Y(u) = \lambda W(u) Y(u), \qquad (2.6.12)$$

where $Y(u) = y(x)$, $Q(u) = p(x)q(x)$ and $W(u) = p(x)w(x)$. We then apply the result just proved for $p(x) = 1$ to (2.6.12), and (2.6.3) follows immediately.

To identify when equality occurs in (2.6.3), we note first that, following the use of the Cauchy-Schwarz inequality, strict inequality occurs in (2.6.6) since y' is not a constant. Hence, comparing (2.6.5) and (2.6.7), equality holds in (2.6.7) only if $q = 0$ a.e. Also, equality can only hold in (2.6.11) when $w = M$ a.e., and this gives rise to the case $pw = M$ a.e. for general p. □

2.6. The upper end-points of the stability intervals

In the particular case when $p = w = 1$ and $n = 1$, (2.6.3) becomes

$$\mu_0 \geq \frac{\pi^2}{a^2}\left(1 + \frac{a}{4}\int_0^a q_-\right) \qquad (2.6.13)$$

with equality only when $q = 0$ a.e.

We next give an example to show that the number 4 which appears in (2.6.3) and (2.6.13) is best possible, that is, it cannot be replaced by a larger constant. Let δ be a real number such that $0 < \delta < a/2$ and let ψ be a real-valued function with a continuous second derivative in $[0, a]$ such that

$$\psi(x) = \begin{cases} x & (0 \leq x \leq \frac{a}{2} - \delta), \\ a - x & (\frac{a}{2} + \delta \leq x \leq a), \end{cases} \qquad (2.6.14)$$

and $\psi > 0$, $\psi'' < 0$ in $(\frac{a}{2} - \delta, \frac{a}{2} + \delta)$. Now define

$$q = \psi''/\psi$$

in $(\frac{a}{2} - \delta, \frac{a}{2} + \delta)$ and $q = 0$ elsewhere in $[0, a]$. Then

$$\psi'' - q\psi = 0$$

in $[0, a]$. Thus ψ satisfies (1.5.2) with $\lambda = 0$, $p = w = 1$ and the above q. By (2.6.14), ψ is also a-semi-periodic and, since ψ has only one zero in $[0, a)$, it follows from Theorem 2.5.1 (b) that ψ is an eigenfunction corresponding to either μ_0 or μ_1. Hence one of μ_0 and μ_1 is zero and, in either case,

$$\mu_0 \leq 0. \qquad (2.6.15)$$

Next,

$$\int_0^a q_- = \int_{\frac{a}{2}-\delta}^{\frac{a}{2}+\delta} \psi''/\psi = [\psi'/\psi]\Big|_{\frac{a}{2}-\delta}^{\frac{a}{2}+\delta} + \int_{\frac{a}{2}-\delta}^{\frac{a}{2}+\delta} (\psi'/\psi)^2$$
$$> (\frac{a}{2} - \delta)^{-1}\left(\psi'(\frac{a}{2} + \delta) - \psi'(\frac{a}{2} - \delta)\right)$$
$$= -4(a - 2\delta)^{-1}.$$

Hence

$$1 + \frac{a}{4(1 - 2\delta/a)^{-1}}\int_0^a q_- > 0.$$

It now follows from (2.6.15) that (2.6.13) does not hold with 4 replaced by the larger constant $4(1 - 2\delta/a)^{-1}$. Since δ can be arbitrarily small, the best-possible nature of 4 in (2.6.13) is demonstrated.

2.7 A step-function example

The ordering (2.4.6) of the periodic and semi-periodic eigenvalues λ_n and μ_n is most simply illustrated by the well-known example of Hill's equation where $p = w = 1$ and $q = 0$. Here we have $\lambda_0 = 0$ and, for $m \geq 0$,

$$\lambda_{2m+1} = \lambda_{2m+2} = 4(m+1)^2\pi^2/a^2,$$
$$\mu_{2m} = \mu_{2m+1} = (2m+1)^2\pi^2/a^2.$$

For a more substantial example of (2.4.6), we take $p = 1$, $q = 0$ and w to be a two-valued step-function

$$w(x) = \begin{cases} w_1 & (0 \leq x < a_1), \\ w_2 & (a_1 \leq x < a), \end{cases}$$

where $w_1 \neq w_2$ and with some a_1 in $(0, a)$. Now (1.5.2) is simply

$$y'' + \lambda w y = 0. \tag{2.7.1}$$

We again have the first periodic eigenvalue $\lambda_0 = 0$. For $\lambda > 0$, (2.7.1) is soluble in terms of sines and cosines on either side of the interface point a_1. Then matching of a solution and its first derivative at the interface gives a solution on the whole of $(0, a)$. In particular, we do this for the solutions ϕ_1 and ϕ_2 forming the canonical fundamental system, i.e., with initial values

$$\phi_1(0, \lambda) = 1, \quad \phi_1'(0, \lambda) = 0; \quad \phi_2(0, \lambda) = 0, \quad \phi_2'(0, \lambda) = 1.$$

We write

$$\nu = \sqrt{\lambda}, \quad A_1 = a_1\sqrt{w_1}, \quad A_2 = (a - a_1)\sqrt{w_2}, \quad \sigma = \sqrt{w_2/w_1}, \tag{2.7.2}$$

and we note that $\sigma \neq 1$. Then, omitting the straightforward details of the calculations, we find that

$$\phi_1(a, \lambda) = \cos \nu A_1 \cos \nu A_2 - \frac{1}{\sigma} \sin \nu A_1 \sin \nu A_2,$$
$$\phi_2'(a, \lambda) = \cos \nu A_1 \cos \nu A_2 - \sigma \sin \nu A_1 \sin \nu A_2.$$

Hence Hill's discriminant is, by (1.5.6),

$$D(\lambda) := \phi_1(a, \lambda) + \phi_2'(a, \lambda)$$
$$= 2\cos \nu A_1 \cos \nu A_2 - \left(\sigma + \frac{1}{\sigma}\right)\sin \nu A_1 \sin \nu A_2$$
$$= \cos \nu I + \cos \nu J + \frac{1}{2}\left(\sigma + \frac{1}{\sigma}\right)(\cos \nu I - \cos \nu J) \tag{2.7.3}$$

2.7. A step-function example

where
$$I = A_1 + A_2, \quad J = |A_1 - A_2|. \tag{2.7.4}$$

Thus the equations $D(\lambda) = \pm 2$ for the λ_n and μ_n become
$$(\sigma+1)^2 \cos \nu I - (\sigma-1)^2 \cos \nu J = \pm 4\sigma. \tag{2.7.5}$$

Explicit solutions of (2.7.5) can be found when there is a suitable connection between I and J, and here we consider the case where
$$I = 2J. \tag{2.7.6}$$

Thus, by (2.7.4), either $A_1 = 3A_2$ or $A_2 = 3A_1$. We take the former, and then (2.7.2) gives
$$\sigma = \frac{1}{3}a_1/(a-a_1). \tag{2.7.7}$$

By (2.7.6), (2.7.5) becomes
$$2\cos^2\theta - \frac{(\sigma-1)^2}{(\sigma+1)^2}\cos\theta - 1 \mp \frac{4\nu}{(\sigma+1)^2} = 0, \tag{2.7.8}$$

where
$$\theta = \nu J \tag{2.7.9}$$

and the $\mp$ produce the λ_n and μ_n respectively.

(a) *The periodic eigenvalues λ_n.* Taking the minus alternative in (2.7.8), we can factorize to obtain
$$(\cos\theta - 1)(\cos\theta - \cos\alpha) = 0,$$

where
$$\alpha = \cos^{-1}\left(-\frac{1}{2} - \frac{2\sigma}{(\sigma+1)^2}\right) \quad \left(\frac{\pi}{2} < \alpha < \pi\right). \tag{2.7.10}$$

The first factor gives
$$\nu J = 2(m+1)\pi \quad (m \in \mathbb{N}_0).$$

These values yield double eigenvalues because, by (2.7.3) and (2.7.6), $D'(\lambda)$ is zero at these values. The second factor gives
$$\nu J = \begin{cases} 2m\pi + \alpha \\ 2(m+1)\pi - \alpha \end{cases} \quad (m \in \mathbb{N}_0).$$

Thus, altogether, we obtain the periodic eigenvalues $\lambda_0 = 0$ and, for $m \geq 0$,
$$\left.\begin{array}{l} \lambda_{4m+1} = (2m\pi + \alpha)^2/J^2, \quad \lambda_{4m+2} = (2(m+1)\pi - \alpha)^2/J^2, \\ \lambda_{4m+3} = \lambda_{4m+4} = 4(m+1)^2\pi^2/J^2. \end{array}\right\} \tag{2.7.11}$$

(b) *The semi-periodic eigenvalues* μ_n. Taking the plus alternative in (2.7.8), we write the equation as
$$\cos^2\theta - \tau\cos\theta - \tau = 0,$$
where
$$\tau = \tfrac{1}{2}(\sigma-1)^2/(\sigma+1)^2 \qquad (2.7.12)$$
and we note that $0 < \tau < \tfrac{1}{2}$. Thus $\cos\theta = \cos\beta$ or $\cos\theta = \cos\gamma$, where
$$\beta = \cos^{-1}\left(\tfrac{1}{2}\left(\tau + \sqrt{\tau^2 + 4\tau}\right)\right), \qquad \gamma = \cos^{-1}\left(\tfrac{1}{2}\left(\tau - \sqrt{\tau^2 + 4\tau}\right)\right) \qquad (2.7.13)$$
and
$$0 < \beta < \gamma < \pi. \qquad (2.7.14)$$
Hence, recalling (2.7.9), we obtain
$$\nu J = \begin{cases} 2m\pi + \beta \\ 2(m+1)\pi - \beta \end{cases} \quad (m \in \mathbb{N}_0)$$
and similarly with γ in place of β. Thus, by (2.7.14), we obtain the semi-periodic eigenvalues
$$\left.\begin{array}{ll}\mu_{4m} = (2m\pi + \beta)^2/J^2, & \mu_{4m+1} = (2m\pi + \gamma)^2/J^2, \\ \mu_{4m+2} = (2(m+1)\pi - \gamma)^2/J^2, & \mu_{4m+3} = (2(m+1)\pi - \beta)^2/J^2.\end{array}\right\} \quad (2.7.15)$$

The ordering (2.4.6) can now be seen in the values of the λ_n and μ_n in (a) and (b) provided that $\gamma < \alpha$. Thus, by (2.7.10) and (2.7.13), we have to show that
$$1 + \frac{4\sigma}{(\sigma+1)^2} > \sqrt{\tau^2 + 4\tau} - \tau,$$
i.e.,
$$2 - 2\tau > \sqrt{\tau^2 + 4\tau} - \tau,$$
by (2.7.12). Now, by (2.7.12) again, $\tau < \tfrac{1}{2}$ giving
$$\sqrt{\tau^2 + 4\tau} + \tau < 2,$$
and we are done.

A simple choice of the parameters in this example is
$$w_1 = 9, \quad w_2 = 1, \quad a_1 = \tfrac{a}{2}. \qquad (2.7.16)$$
It is easy to check that $J = a$ in (2.7.4) and that (2.7.6) holds. Then (2.7.7) and (2.7.12) give
$$\sigma = \tfrac{1}{3}, \quad \tau = \tfrac{1}{8}$$
and finally (2.7.10) and (2.7.13) give
$$\alpha = \cos^{-1}(-\tfrac{7}{8}), \quad \beta = \cos^{-1}(\tfrac{1+\sqrt{33}}{16}), \quad \gamma = \cos^{-1}(\tfrac{1-\sqrt{33}}{16}).$$

2.8 Even coefficients

In the case where p, q and w are even functions in (1.5.2), there are four eigenvalue problems, each with separated boundary conditions, whose eigenvalues and eigenfunctions are usefully related to those for the periodic and semi-periodic problems over $[0, a]$. These four problems comprise (1.5.2) on the half-interval $[0, \frac{1}{2}a]$ together with (in turn) the boundary conditions:

$$
\begin{aligned}
&\text{(I)} && y'(0) = y'(\tfrac{1}{2}a) = 0 \\
&\text{(II)} && y(0) = y(\tfrac{1}{2}a) = 0 \\
&\text{(III)} && y'(0) = y(\tfrac{1}{2}a) = 0 \\
&\text{(IV)} && y(0) = y'(\tfrac{1}{2}a) = 0.
\end{aligned}
$$

We denote the eigenvalues and eigenfunctions in these problems by λ_n^I, ψ_n^I etc, and their relation to the periodic λ_n, ψ_n and semi-periodic μ_n, ξ_n is given in the two propositions which follow. For the proofs, we recall the basic oscillation property that the n-th eigenfunction (such as ψ_n^I etc) in problems with separated boundary conditions has exactly n zeros in the open interval in question, here $(0, \frac{1}{2}a)$ (see Theorem 2.3.4 (a)).

Theorem 2.8.1. (a) *For $n \geq 1$, λ_n^I and λ_{n-1}^{II} are the same as λ_{2n-1} and λ_{2n} (but not necessarily in that order). Also, ψ_n^I is even and ψ_{n-1}^{II} is odd. For $n = 0$, $\lambda_0^I = \lambda_0$ and ψ_0^I is even.*

(b) *For $n \geq 0$, λ_n^{III} and λ_n^{IV} are the same as μ_{2n} and μ_{2n+1} (but not necessarily in that order). Also, ψ_n^{III} is even and ψ_n^{IV} is odd.*

Proof. (a) Since p, q and w are all even, $y(-x)$ is also a solution of (1.5.2) when $y(x)$ is. For Problem I, the initial condition $y'(0) = 0$ means that ψ_n^I is even:

$$\psi_n^I(-x) = \psi_n^I(x),$$

these two solutions of (1.5.2) having the same initial values at $x = 0$. The other condition $y'(\frac{1}{2}a) = 0$ then shows that the derivative of ψ_n^I is zero at both $-\frac{1}{2}a$ and $\frac{1}{2}a$. Thus ψ_n^I has period a. Further, ψ_n^I has exactly n zeros in the open interval $(0, \frac{1}{2}a)$ and hence exactly $2n$ zeros in $[-\frac{1}{2}a, \frac{1}{2}a)$. Hence, by Theorem 2.5.1 (a), λ_n^I is either λ_{2n-1} or λ_{2n}. The rest of part (a) is proved similarly, where we note that ψ_{n-1}^{II} also has $2n$ zeros in $[-\frac{1}{2}a, \frac{1}{2}a)$: $2(n-1)$ of them in $(-\frac{1}{2}a, 0) \cup (0, \frac{1}{2}a)$ and the other 2 at $-\frac{1}{2}a$ and 0.

The proof of part (b) is again similar, making use of Theorem 2.5.1 (b). □

This theorem will be used in a further discussion of the Mathieu equation in section 3.7 but, meanwhile, a more explicit illustration of it is provided by two examples.

Example 1. Let $p = 1$, $q = 0$ and let w be the step-function

$$w(x) = \begin{cases} 9 & (0 \leq x < \tfrac{1}{4}a, \quad \tfrac{3}{4}a \leq x < a), \\ 1 & (\tfrac{1}{4}a \leq x < \tfrac{3}{4}a), \end{cases} \tag{2.8.1}$$

a-periodically extended to obtain an even function on $\mathbb{R}$. This example is the same as in (2.7.16) with x shifted through $\tfrac{1}{4}a$. This does not affect the values of the periodic and semi-periodic eigenvalues found above in (2.7.11) and (2.7.15).

Referring in particular to Proposition 2.8.1 (b), a simple but careful calculation shows that, for (2.8.1),

$$(\mu_{8m}, \mu_{8m+1}) = (\lambda_{4m}^{\text{III}}, \lambda_{4m}^{\text{IV}}), \qquad (\mu_{8m+2}, \mu_{8m+3}) = (\lambda_{4m+1}^{\text{III}}, \lambda_{4m+1}^{\text{IV}}),$$

$$(\mu_{8m+4}, \mu_{8m+5}) = (\lambda_{4m+2}^{\text{IV}}, \lambda_{4m+2}^{\text{III}}), \quad (\mu_{8m+6}, \mu_{8m+7}) = (\lambda_{4m+3}^{\text{IV}}, \lambda_{4m+3}^{\text{III}})$$

as ordered pairs. Thus sometimes III comes before IV and sometimes vice versa.

Example 2. Let $p = w = 1$ and

$$q(x) = c \cos 2x - 2\alpha^2 \cos 4x$$

where c and α are non-zero constants. Then (1.5.2) has the special feature that the transformation

$$y = z \exp(\alpha \cos 2x)$$

produces a z-equation in which the argument $4x$ does not appear. Thus

$$z''(x) - 4\alpha \sin 2x \, z'(x) + (\lambda + 2\alpha^2 - (c + 4\alpha) \cos 2x) \, z(x) = 0.$$

It is now simple to verify the following explicit solutions of the z-equation with the corresponding values of λ. In (i) "even" comes before "odd" and, in (ii), "odd" comes before "even".

(i) Let $c = -8\alpha$. Then there are solutions

$$\cos x, \quad \lambda = 1 - 4\alpha - 2\alpha^2,$$

$$\sin x, \quad \lambda = 1 + 4\alpha - 2\alpha^2.$$

By Theorem 2.5.1 (b), these eigenvalues are μ_0 and μ_1.

(ii) Let $c = -12\alpha$. Then there are solutions

$$\sin 2x, \quad \lambda = 4 - 2\alpha^2,$$

$$\cos 2x + \left(1 - \sqrt{1 + 16\alpha^2}\right)/4\alpha, \quad \lambda = 2 + 2\sqrt{1 + 16\alpha^2} - 2\alpha^2.$$

By Theorem 2.5.1 (a), these eigenvalues are λ_1 and λ_2.

2.9 Comparison of eigenvalues

In this section we compare the eigenvalues in two separate boundary-value problems. For the Sturm-Liouville case (1.5.2), the coefficients p, q, w and p_1, q_1, w_1 in the two problems satisfy

$$p_1 \geq p, \quad q_1 \geq q, \quad 0 < w_1 \leq w \tag{2.9.1}$$

in $[0, a]$. For the Dirac case (1.5.4),

$$q_1 \geq q \tag{2.9.2}$$

but the p_1 and p_2 coefficients remain the same in the two problems. It is a question of how the corresponding eigenvalues compare when the coefficients are related as in (2.9.1) and (2.9.2).

We consider first the easiest situation where the boundary conditions are separated as in (2.3.2), and we can also add the further distinguishing features

$$0 \leq \alpha_1 \leq \alpha \quad \pi \geq \beta_1 \geq \beta \tag{2.9.3}$$

between the two problems.

Theorem 2.9.1. (a) *For the Sturm-Liouville problem on $[0, a]$ with separated boundary conditions, let (2.9.1) and (2.9.3) hold. Then the eigenvalues λ_n and $\lambda_{1,n}$ in the two problems satisfy*

1. *if $w_1 = w$ a.e., then $\lambda_{1,n} \geq \lambda_n$ for $n \geq 0$;*
2. *otherwise, $\lambda_{1,n} \geq \lambda_n$ provided that $\lambda_n \geq 0$.*

(b) *For the Dirac problem on $[0, a]$ with separated boundary conditions, let (2.9.2) and (2.9.3) hold. Then $\lambda_{1,n} \geq \lambda_n$ for $n \in \mathbb{Z}$.*

Proof. For both (a) and (b), let $\theta(x, \lambda)$ and $\theta_1(x, \lambda)$ be the Prüfer angles for the two problems being compared, respectively. Then

$$\theta_1(a, \lambda) = \alpha_1 \leq \alpha = \theta(a, \lambda).$$

Then, under the stated conditions, we can apply Theorem 2.3.1 to the Prüfer equations (2.2.5) and (2.2.7) for the two problems, to obtain $\theta_1(x, \lambda) \leq \theta(x, \lambda)$ in $[0, a]$ and, in particular,

$$\theta_1(a, \lambda_n) \leq \theta(a, \lambda_n) = \beta + n\pi$$

as in Theorem 2.3.4. Thus $\theta_1(a, \lambda_n) \leq \beta_1 + n\pi = \theta_1(a, \lambda_{1,n})$, giving $\lambda_{1,n} \geq \lambda_n$ by the monotonicity of θ_1 as a function of λ (see Theorem 2.3.3). □

Moving on to the boundary conditions (1.8.1), the situation becomes less simple because we no longer have the initial condition $\theta(0, \lambda) = \alpha$, and Theorem

2.3.1 cannot be applied. There is however an alternative method for the Sturm-Liouville case which is based on the Dirichlet integral

$$J(f,g) := \int_0^a (pf'\bar{g}' + qf\bar{g}) \tag{2.9.4}$$

defined for $AC[0,a]$ functions f and g. We first develop the properties of $J(f,g)$ which we need, bearing in mind that the boundary conditions under consideration now are (1.8.2), to include the periodic, semi-periodic and ω-twisted cases.

If, in addition, g' is $AC[0,a]$, an integration by parts in (2.9.4) gives

$$J(f,g) = -\int_0^a f\{(p\bar{g}')' - q\bar{g}\} + [pf\bar{g}']\Big|_0^a. \tag{2.9.5}$$

If f satisfies the first, and g the second, boundary condition in (1.8.2), the integrated terms cancel out since $|\omega| = 1$. In particular, if g is an eigenfunction ψ_n for (1.5.2) and (1.8.2) corresponding to the eigenvalue λ_n, (2.9.5) gives

$$J(f,\psi_n) = \lambda_n c_n, \tag{2.9.6}$$

where c_n denotes the Fourier coefficient

$$c_n := \int_0^a f(x)\overline{\psi}_n(x)w(x)\,dx \tag{2.9.7}$$

and we recall that λ_n is real, by Proposition 1.8.1. If now ψ_n is taken to be normalised and we also recall the orthogonality property (1.8.4), then a special case of (2.9.6) is

$$J(\psi_m, \psi_n) = \begin{cases} \lambda_n & (m=n), \\ 0 & (m \neq n). \end{cases} \tag{2.9.8}$$

Theorem 2.9.2. *Let f be $AC[0,a]$ and satisfy the first boundary condition in (1.8.2). Then, with the Fourier coefficients c_n defined as above,*

$$\sum_{n=0}^{\infty} \lambda_n |c_n|^2 \leq J(f,f). \tag{2.9.9}$$

Proof. We suppose first that $q \geq 0$. Then, by (2.9.4), $J(g,g) \geq 0$ and, in particular,

$$J\left(f - \sum_{n=0}^N c_n \psi_n, f - \sum_{n=0}^N c_n \psi_n\right) \geq 0$$

where $N \in \mathbb{N}$, On 'multiplying out' the left-hand side here, we obtain

$$J(f,f) - \sum_{n=0}^N c_n J(\psi_n, f) - \sum_{n=0}^N \bar{c}_n J(f, \psi_n) + \sum_{n=0}^N \lambda_n c_n \bar{c}_n \geq 0,$$

2.9. Comparison of eigenvalues

where we have used (2.9.8) to deduce the last summation. Since $J(\psi_n, f) = \overline{J(f, \psi_n)}$, we obtain

$$\sum_{n=0}^{N} \lambda_n |c_n|^2 \leq J(f, f),$$

on using (2.9.6). Then (2.9.9) follows on letting $N \to \infty$.

To prove (2.9.9) without the assumption that $q \geq 0$, let K be a constant which is sufficiently large to make $q + Kw \geq 0$ in $[0, a]$. Then (1.5.2) can be written as

$$-(py')' + Qy = \Lambda wy,$$

where $\Lambda = \lambda + K$ and $Q = q + Kw$. Since $Q \geq 0$, the above proof gives

$$\sum_{n=0}^{\infty} (\lambda_n + K)|c_n|^2 \leq \int_0^a \left(p|f'|^2 + (q + Kw)|f|^2 \right).$$

The terms involving K on each side are equal on account of the Parseval formula

$$\sum_{n=0}^{\infty} |c_n|^2 = \int_0^\infty |f|^2 w \qquad (2.9.10)$$

(for which we refer forward to (4.2.8)). Now (2.9.9) follows in the general case. $\square$

We are now in a position to prove the result which corresponds to Theorem 2.9.1(a).

Theorem 2.9.3. *For the Sturm-Liouville problem on $[0, a]$ with the boundary condition (1.8.2), let (2.9.1) hold. Then parts (i) and (ii) of Theorem 2.9.1 (a) continue to hold.*

Proof. Let $\psi_{1,n}$ denote the normalised eigenfunctions corresponding to $\lambda_{1,n}$ and let $J_1(f, g)$ be the Dirichlet integral (2.9.4) with p and q replaced by p_1 and q_1. By (2.9.4) we have

$$J_1(f, f) \geq J(f, f). \qquad (2.9.11)$$

Now consider the choice

$$f = \gamma_0 \psi_{1,0} + \cdots + \gamma_m \psi_{1,m}, \qquad (2.9.12)$$

where the constants γ_i are chosen so that

$$|\gamma_0|^2 + \cdots + |\gamma_m|^2 = 1 \qquad (2.9.13)$$

and the first m Fourier coefficients of f are zero, that is,

$$c_n = 0 \quad (0 \leq n \leq m - 1) \qquad (2.9.14)$$

with the c_n defined as in (2.9.7). Thus, by (2.9.7) and (2.9.12), we have m homogeneous linear algebraic equations (2.9.14) to be satisfied by the $m+1$ numbers $\gamma_0, \ldots, \gamma_m$, and such numbers aways exist satisfying the normalising condition (2.9.13). We note that (2.9.13) and the orthogonality of the $\psi_{1,n}$ give

$$\int_0^a |f|^2 w_1 = 1. \tag{2.9.15}$$

By (2.9.9) and (2.9.14),

$$J(f,f) \geq \sum_m^\infty \lambda_n |c_n|^2 \geq \lambda_m \sum_m^\infty |c_n|^2 = \lambda_m \int_0^a |f|^2 w \tag{2.9.16}$$

by the Parseval formula (2.9.10). Also, by (2.9.8) applied to J_1,

$$J_1(f,f) = \lambda_{1,0}|\gamma_0|^2 + \cdots + \lambda_{1,m}|\gamma_m|^2 \leq \lambda_{1,m}(|\gamma_0|^2 + \cdots + |\gamma_m|^2).$$

Hence, by (2.9.11), (2.9.13), (2.9.15) and (2.9.16), we have

$$\lambda_{1,m} \geq \lambda_m \int_0^a |f|^2 w \geq \lambda_m \int_0^a |f|^2 w_1 = \lambda_m$$

provided that either $w = w_1$ or $\lambda_m \geq 0$ (or both). Thus $\lambda_{1,m} \geq \lambda_m$, as required. $\square$

2.10 Least eigenvalues

There is an important consequence of (2.9.9) concerning the least eigenvalue λ_0. From (2.9.9) and (2.9.10), we have

$$J(f,f) \geq \lambda_0 \sum_{n=0}^\infty |c_n|^2 = \lambda_0 \int_0^a |f|^2 w,$$

and the inequality is strict unless $c_n = 0$ for all n such that $\lambda_n > \lambda_0$, that is, unless f is an eigenfunction corresponding to λ_0. Thus

$$\lambda_0 = \min\left(J(f,f) \Big/ \int_0^a |f|^2 w\right), \tag{2.10.1}$$

the minimum being taken over all non-trivial f in $AC[0, a]$ which satisfy the first boundary condition $f(a) = \omega f(0)$ in (1.8.2).

In this section we make two applications of (2.10.1) to derive inequalities for λ_0 in terms of the coefficients in (1.5.2). In the first theorem we write $\lambda_0(\omega)$ to emphasise that the boundary conditions are (1.8.2).

2.10. Least eigenvalues

Theorem 2.10.1. *Let $\lambda_0(w)$ be the least eigenvalue in the boundary-value problem (1.5.2), (1.8.2). Then*

$$\lambda_0(1) \leq \lambda_0(\omega) \leq \lambda_0(1) + \left(\frac{\arg \omega}{a}\right)^2 \frac{\sup p}{\inf w}. \tag{2.10.2}$$

Proof. Here $\lambda_0(1)$ is the least periodic eigenvalue over $[0,a]$, and the left-hand inequality follows from (1.8.5) since $\lambda_0(1)$ is the least point of the stability set S. To prove the right-hand inequality in (2.10.2), let ψ_0 be the (real-valued) eigenfunction corresponding to $\lambda_0(1)$, and define

$$f(x) = \psi_0(x)\omega^{x/a}.$$

Then f satisfies (1.8.2). Also $|f| = |\psi_0|$ and

$$J(f, f) = \int_0^a \left(p|\psi_0' + (a^{-1}\log \omega)\psi_0|^2 + q\psi_0^2 \right)$$

$$= J(\psi_0, \psi_0) + \left(\frac{\arg \omega}{a}\right)^2 \int_0^a p\psi_0^2$$

since $|\omega| = 1$. Using this in (2.10.1), we obtain

$$\lambda_0(\omega) \leq \lambda_0(1) + \left(\frac{\arg \omega}{a}\right)^2 \int_0^a p\psi_0^2 \Big/ \int_0^a \psi_0^2 w, \tag{2.10.3}$$

and the right-hand inequality in (2.10.2) follows. □

Since $\lambda_0(-1)$ is the first semi-periodic eigenvalue over $[0, a]$, we obtain the following corollary by taking $\omega = -1$ in (2.10.2).

Corollary 2.10.2. *The length of the first stability interval does not exceed*

$$(\pi^2 \sup p)/(a^2 \inf w).$$

Another consequence of (2.10.3) is that, when $q(x) = 0$,

$$0 \leq \lambda_0(\omega) \leq \left(\frac{\arg \omega}{a}\right)^2 \int_0^a p \Big/ \int_0^a w.$$

This follows from (2.10.3) since now $\lambda_0(1) = 0$ and $\psi_0(x) = a^{-\frac{1}{2}}$.

The next theorem concerns the least periodic eigenvalue $\lambda_0(1)$, and it complements Theorem 2.5.3 by providing a lower estimate for $\lambda_0(1)$. We define the constant d by

$$d = \int_0^a q \Big/ \int_0^a w. \tag{2.10.4}$$

Theorem 2.10.3. Let $L = \inf_{[0,a]}(pw)$. Then

$$d - \frac{1}{16L}\left(\int_0^a |q - dw|\right)^2 \leq \lambda_0(1) \leq d, \qquad (2.10.5)$$

and the number 16 is best possible.

Proof. The right-hand inequality was proved in Theorem 2.5.3. To prove the left-hand inequality, let f and g be real valued, a-periodic $AC[0,a]$ functions. Then, by (2.9.4)

$$J(f,f) = \int_0^a \left(p(f' + gf)^2 - 2pgf'f + (q - pg^2)f^2\right). \qquad (2.10.6)$$

But

$$\int_0^a 2pgf'f = -\int_0^a (pg)'f^2$$

on integrating by parts and using the periodicity of pgf^2. Hence (2.10.6) gives

$$J(f,f) \geq \int_0^a ((pg)' + q - pg^2)f^2$$

$$\geq \left(\inf_{[0,a]}((pg)' + q - pg^2)/w\right)\int_0^a f^2 w.$$

It now follows from (2.10.1) that

$$\lambda_0(1) \geq \inf_{[0,a]}((pg)' + q - pg^2)/w. \qquad (2.10.7)$$

We have now to make a choice for g, and we choose $g = -(g_1 - k)/p$, where

$$g_1(x) = \int_0^x (q - dw) \qquad (2.10.8)$$

and k is a constant yet to be specified. By (2.10.4), g_1 has period a and hence (2.10.7) gives

$$\lambda_0(1) \geq d + \inf_{[0,a]}(-pg^2/w) \geq d - \frac{1}{L}\sup_{[0,a]}|g_1 - k|^2. \qquad (2.10.9)$$

Now the choice $k = 0$ would yield (2.10.5) but without the factor 16. Thus, to prove (2.10.5) as stated, we make a less obvious choice, and we proceed as follows. Since g_1 is continuous and a-periodic, there are points X and Y at which g_1 attains its infimum and supremum respectively, and such that $X \leq Y < X + a$. We now choose k so that

$$g_1(Y) - k = -(g_1(X) - k),$$

2.10. Least eigenvalues

that is
$$k = \tfrac{1}{2}(g_1(X) + g_1(Y)).$$

Then
$$\sup_{[0,a]} |g_1 - k| = g_1(Y) - k = \tfrac{1}{2}(g_1(Y) - g_1(X)). \tag{2.10.10}$$

Writing $q_1 = q - dw$, we have from (2.10.8)
$$g_1(Y) - g_1(X) = \int_X^Y q_1 \leq \int_X^Y |q_1|. \tag{2.10.11}$$

Also, since g_1 has period a,
$$\int_X^Y q_1 = -\int_Y^{X+a} q_1.$$

Hence again
$$g_1(Y) - g_1(X) \leq \int_Y^{X+a} |q_1|. \tag{2.10.12}$$

Adding (2.10.11) and (2.10.12), we obtain
$$2(g_1(Y) - g_1(X)) \leq \int_X^{X+a} |q_1| = \int_0^a |q_1|.$$

Now (2.10.5) follows from this and (2.10.9) and (2.10.10).

Finally, to show that 16 is best possible, we give an example in which $p = w = 1$. We introduce constants δ and η, where $\delta > 0$ and $0 < \eta < 1$, to be chosen later. Let $a = 2\delta + 2$ and consider
$$q(x) = \begin{cases} -1 & (0 \leq x \leq \delta), \\ 1 & (\delta + 1 < x \leq 2\delta + 1), \end{cases}$$

while $q(x) = 0$ elsewhere in $(\delta, 2\delta + 2)$. Then, first of all, $d = 0$ in (2.10.4) and (2.10.5). Next, in (2.10.1), we take f to be the function whose graph in the plane $\mathbb{R}^2$ consists of straight line segments joining the points $(0, 1)$, $(\delta, 1)$, $(\delta + 1, 1 - \eta)$, $(2\delta + 1, 1 - \eta)$ and $(2\delta + 2, 1)$. A calculation gives
$$\lambda_0 / \left(\int_0^a |q|\right)^2 \leq \frac{\tfrac{1}{4}(2\eta^2/\delta^2 - (2\eta - \eta^2)/\delta)}{\delta(2 + 2\eta + \eta^2) + 2(1 - \eta + \tfrac{1}{3}\eta^2)}. \tag{2.10.13}$$

We minimise the numerator on the right-hand side, considered as a quadratic in $1/\delta$, by choosing $\delta = 4\eta/(2 - \eta)$. Then the right-hand side of (2.10.13) becomes
$$\frac{-\tfrac{1}{32}(2 - \eta)^2}{\delta(2 + 2\eta + \eta^2) + 2(1 - \eta + \tfrac{1}{3}\eta^2)}. \tag{2.10.14}$$

If η were zero, (2.10.14) would have the value $-1/16$. Hence, by choice of $\eta \in (0, 1)$ and therefore of $\delta > 0$, we can make (2.10.14) arbitrarily close to $-1/16$, as required. $\square$

2.11 Chapter notes

§2.2 The Prüfer transformation appeared first in [144]. Rofe-Beketov has an extension of this idea to higher-order systems, see [156].

The Kepler transformation [161] is named after Kepler's method of calculating the area of an ellipse sector by deforming the ellipse into a circle.

§2.3 We refer to Walter [192] for the theory of differential inequalities. The Lipschitz condition in Theorem 2.3.1 can be relaxed to the one-sided condition

$$f(x,y) - f(x,z) \le K(x)(y-z) \quad (y \ge z).$$

The generalised Sturm comparison theorem, Corollary 2.3.2, was given by Weidmann [194, Theorem 16.1].

§2.4 The rotation number theory has also been developed for more general spectral problems involving almost periodic potentials by Johnson and Moser [103], and for the p-Laplacian by Zhang [209]. In the latter case however, there can be additional periodic eigenvalues lying between those arising from the rotation number method; see Binding and Rynne [14].

The quasi-momentum extends to an analytic function on $\mathbb{C} \setminus \overline{\mathcal{I}}$, giving a conformal mapping between this set and the slitted complex plane with symmetric vertical cuts of different heights at $\operatorname{Re} z = n\pi$ with non-zero integer n. The complex quasi-momentum plays an important role in the inverse theory of the periodic equation [128]. Details can be found in [116].

The Sturm-Liouville part of Theorem 2.4.2 has a long history: Coddington and Levinson [28, Chapter 8, Theorem 3.1], Ince [97, §10.8], Magnus and Winkler [127, Chapter 2], Titchmarsh [183, §21.4]. These results go back to Lyapunov (1899) in the case of $q = 0$– Starzhinskii [173]– and to Hamel [71] and Haupt [80]; see also Kramers [118].

In the Sturm-Liouville case, the interlacing property (2.4.6) also holds for more general coupled boundary conditions [52, 112]. An early reference is Birkhoff [15].

For the general system (1.5.1) with W positive definite, (2.4.6) is due to Harris [77, Theorem 2.6] with a proof based on the discriminant $D(\lambda)$; see also Yakubovich and Starzhinskii [203, Chapter VIII, §6].

The inequality in Theorem 2.4.4 was shown for the one-dimensional periodic Schrödinger operator ($w = p = 1$) by Moser [137].

The Dirac part of Theorem 2.4.5 is new here. For the Sturm-Liouville part, see also [48, Theorem 3.1.3] and [90]. An early result related to the rotation number is in Wallach [191, Theorem II].

§2.5 The first part of Theorem 2.5.2 is in [127, Chapter 4]. Our proof here appears to be new. Another proof is in [48, Theorem 3.2.1]. The second part of Theorem 2.5.2 is due to Moore [136]; see also Swanson [179, Chapter 2, §11].

2.11. Chapter notes

Theorem 2.5.3 is due to Borg [17]. The proof here follows Ungar [189]. Given the properties of $J(f, f)$ in sections 2.9 and 2.10 below, another proof of the theorem simply takes $f = 1$ in (2.10.1).

Theorem 2.5.3 is due to Blumenson [16]; see also [48, §3.3].

§2.6 Putnam [147] and [148]. Other inequalities due to Putnam are in [145].

For the Wirtinger inequality (2.6.10), of which there are several proofs, see [72, §7.7]. One proof uses our (2.10.1) as applied to the Dirichlet boundary-value problem on $[c, d]$.

The best possible nature of the number 4 was first proved by van Kampen and Wintner [190]. The proof given here is based on Borg [18].

§2.7 This step-function example was considered by Hochstadt [89] in relation to Theorem 2.4.2 and coexistence. It was also considered by Brown and Eastham [24] in the context of an eigenvalue-inducing interface problem. See also Pipes [142, section III] and, for a related inverse problem, Efendiev and Orudzhev [54].

Another Meissner-type equation $y'' + (\lambda - q)y = 0$, with a three-valued step-function q, was discussed by Yoshitomo [205] where it was shown *inter alia* that the first and second instability intervals are non-vanishing. Other, more specialised papers are [60] and [210] by Gan and Zhang.

§2.8 The four eigenvalue problems were related to the periodic and semi-periodic problems by Hochstadt [85]; see also [48, Theorem 1.3.4].

With coefficients as in Example 2, (1.5.2) is known as the Whittaker-Hill equation [5, p.145, footnote]. The transformation from y to z goes back to Ince [96] and, for a recent use, see Djakov and Mityagin [37, 38] and Hemery and Veselov [82].

§2.9 The form $J(f, f)$ appears to have been first applied to the ω-twisted problem by Strutt [176, 177]. Our treatment of $J(f, f)$ in Theorems 2.9.2 and 2.9.3 follows that of Titchmarsh [183, §§14.1-14.7].

§2.10 Theorem 2.10.1 is due to Eastham [45], [48, Theorem 5.5.3], and a corresponding result for the Schrödinger equation in two or more dimensions is in [46] and [48, Theorem 6.9.3]. Further, more detailed estimates for $\lambda_0(\omega)$ are in Atkinson [6] and Heil [81, Theorem 6.9.3], especially in relation to the weight function w.

Theorem 2.10.3 is due to Kato [107] in the case $p = w = 1$, and it improves on an earlier result of Wintner [202] which had 1 in place of 16 in (2.10.5). Further inequalities of the type (2.10.5) are given by Müller-Pfeiffer [139] and [140, pp. 120-126]. In particular, when $p = w = 1$ and $q \geq 0$, $\pi^2 d/(\pi^2 + a^2 d) \leq \lambda_0(1) \leq d$.

The example showing that 16 is best possible in (2.10.5) is due to Eastham [47].

Chapter 3

Asymptotics

3.1 Introduction

The main purpose of this chapter is to examine the nature of the instability intervals–first their asymptotic lengths as they recede to infinity, and second the more specialised situation when all but a finite number of them are absent. To deal with the lengths we require asymptotic estimates for the eigenvalues λ_n and μ_n as $n \to \infty$, and the theory of Chapters 1 and 2 provides two methods to produce these estimates. Our method of choice in this chapter is based on the Prüfer transformation and oscillation theory of Chapter 2, and the other method (which we shall also touch on) uses a direct examination of the discriminant $D(\lambda)$ as $\lambda \to \infty$. A feature of the asymptotic estimates is that they become increasingly accurate the more times that the coefficients $p(x)$, $q(x)$ and $w(x)$ are differentiable, and we develop this theme in sections 3.3-3.7.

The asymptotic forms of the solutions $\phi_1(x, \lambda)$ and $\phi_2(x, \lambda)$ would be relevant to the $D(\lambda)$ method, but we also need them later in the chapter (sections 3.9-3.11) for our second stated purpose about the absence of instability intervals. These asymptotics are the subject of section 3.8. In the concluding section 3.12, we show that the absence of any instability interval represents a specialised situation, that is, all such intervals are non-empty for most (in a certain sense) coefficients.

3.2 Prüfer transformation formulae

The Prüfer transformation

$$y = \rho \sin \theta, \quad py' = \rho \cos \theta \qquad (3.2.1)$$

was introduced in (2.2.4) for solutions y of

$$(py')' + (\lambda w - q)y = 0. \qquad (3.2.2)$$

In this chapter we require two variations of (3.2.1) but, before introducing them, we note a couple of simplifications of (3.2.2) in which p can be replaced by unity. Without any additional conditions on p, the change of variable from x to t defined by $dt = dx/p(x)$ changes (3.2.2) into

$$y'' + (\lambda w - q)\, y = 0 \qquad (3.2.3)$$

with a new w and q, and the prime now denotes t–differentiation. Again, the Liouville-Green transformation

$$t = \int_0^x (w/p)^{1/2}\, du, \qquad z(t) = (pw)^{1/4}(x)\, y(x)$$

changes (3.2.2) into

$$d^2 z/dt^2 + (\lambda - Q(t))\, z = 0,$$

where $Q(t) = q(x) - (p^{1/4} w^{-3/4})(x) \dfrac{d}{dx} p(x) \dfrac{d}{dx} (pw)^{-1/4}(x)$, subject now to p and w being twice differentiable. Thus any result for (3.2.2) can be obtained from the simpler equation

$$y'' + (\lambda - q)\, y = 0 \qquad (3.2.4)$$

in which both p and w are replaced by unity.

The first variation of the Prüfer transformation, which we apply to (3.2.3), is

$$y = \rho \sin\theta, \qquad y' = \lambda^{1/2} f \rho \cos\theta, \qquad (3.2.5)$$

where $f\ (>0)$ is a differentiable function which is at our choice. As in section 2.2, the corresponding first-order differential equations for θ and ρ are

$$\theta' = \lambda^{1/2}\left(f \cos^2\theta + \frac{w}{f}\sin^2\theta\right) + \tfrac{1}{2}\frac{f'}{f}\sin 2\theta - \lambda^{-1/2}\frac{q}{f}\sin^2\theta, \qquad (3.2.6)$$

$$\rho'/\rho = \tfrac{1}{2}\lambda^{1/2}\left(f - \frac{\lambda w - q}{\lambda f}\right)\sin 2\theta - \frac{f'}{f}\cos^2\theta. \qquad (3.2.7)$$

The second variation, which we apply to (3.2.4), requires q to be differentiable, and it is

$$y = (\lambda - q)^{-1/4}\rho \sin\theta, \qquad y' = (\lambda - q)^{1/4}\rho \cos\theta. \qquad (3.2.8)$$

This time we have

$$\theta' = (\lambda - q)^{1/2} - \frac{1}{4}\frac{q'}{\lambda - q}\sin 2\theta, \qquad (3.2.9)$$

$$\rho'/\rho = \frac{1}{4}\frac{q'}{\lambda - q}\cos 2\theta. \qquad (3.2.10)$$

3.2. Prüfer transformation formulae

Actually, (3.2.8) is the case $f = \lambda^{-1/2}(\lambda - q)^{1/2}$ of (3.2.5) together with a redefinition of ρ, and both are examples of the Kepler transformation in Theorem 2.3.3.

As in Theorem 2.4.2, we continue to have

$$\theta(a,\lambda) - \theta(0,\lambda) = \begin{cases} 2(m+1)\pi & \text{when } \lambda = \lambda_{2m+1}, \lambda_{2m+2}, \\ (2m+1)\pi & \text{when } \lambda = \mu_{2m}, \mu_{2m+1}, \end{cases} \quad (3.2.11)$$

for both (3.2.5) and (3.2.8).

In our use of Prüfer transformations, we need a result which is similar to the classical Riemann-Lebesgue Lemma and which we state and prove now in a general form to cover all our requirements. We are concerned with a fixed x-interval $[c,d]$, $\lambda > \Lambda > 0$ and a function $\Theta(x,\lambda)$ which, as for example $\theta(x,\lambda)$ in (3.2.9), satisfies a differential equation

$$\Theta'(x,\lambda) = \lambda^A F(x) + \lambda^B G(x,\lambda,\Theta(x,\lambda)). \quad (3.2.12)$$

Lemma 3.2.1. *Let F be $AC[c,d]$, $A > 0$, $B < A$,*

$$|F| \geq k > 0, \qquad |G| \leq K \quad (3.2.13)$$

where k and K are constants. Then, if g is $L(c,d)$ and C is a non-zero constant,

$$\int_c^d g(x) \sin(C\Theta(x,\lambda))\,dx \to 0$$

as $\lambda \to \infty$. The same is also true with $\cos$ in place of $\sin$.

Proof. Corresponding to any ϵ (> 0), let $h(x)$ be an absolutely continuous function such that

$$\int_c^d |g(x) - h(x)|\,dx < \epsilon.$$

Then

$$\left| \int_c^d g(x)\sin(C\Theta(x,\lambda))\,dx \right| < \epsilon + \left| \int_c^d h(x)\sin(C\Theta(x,\lambda))\,dx \right|. \quad (3.2.14)$$

By (3.2.12) and (3.2.13),

$$\int_c^d h(x)\sin(C\Theta(x,\lambda))\,dx$$

$$= \lambda^{-A} \int_c^d (h/F)(x)\sin(C\Theta(x,\lambda))\,\Theta'(x,\lambda)\,dx + O(\lambda^{-(A-B)})$$

$$= \lambda^{-A} C^{-1}[-(h/F)(x)\cos(C\Theta(x,\lambda))]\Big|_c^d$$

$$\quad + \lambda^{-A} C^{-1} \int_c^d (h/F)'(x)\cos(C\Theta(x,\lambda))\,dx + O(\lambda^{-(A-B)}).$$

Hence
$$\left|\int_c^d h(x)\sin(C\Theta(x,\lambda))\,dx\right| \le \lambda^{-A}M_1(\epsilon) + \lambda^{-(A-B)}M_2(\epsilon) < \epsilon$$

if λ is large enough, $M_1(\epsilon)$ and $M_2(\epsilon)$ being independent of λ. The lemma now follows from (3.2.14). $\square$

3.3 The coefficient w

The first theorem in this section gives an initial result for λ_n and μ_n where there is no assumption about the differentiability of w and q in (3.2.3). The main focus of the proof is on w and the choice of a suitable f in (3.2.5).

Theorem 3.3.1. *For (3.2.3), λ_{2m+1} and λ_{2m+2} both satisfy*
$$\sqrt{\lambda} = 2(m+1)\pi I^{-1} + o(m) \tag{3.3.1}$$

as $m \to \infty$, where
$$I = \int_0^a w^{1/2}(x)\,dx.$$

The same result holds for μ_{2m} and μ_{2m+1}, but with $2(m+1)$ replaced by $2m+1$.

Proof. We apply the Prüfer transformation (3.2.5) to (3.2.3). By (3.2.6) and (3.2.11) we have for λ_{2m+1} and λ_{2m+2}

$$\begin{aligned}2(m+1)\pi =& \lambda^{1/2}I + \lambda^{1/2}\int_0^a (f - w^{1/2})\,dx + \lambda^{1/2}\int_0^a \left(\frac{w}{f} - f\right)\sin^2\theta\,dx \\ &+ \frac{1}{2}\int_0^a \frac{f'}{f}\sin 2\theta\,dx - \lambda^{-1/2}\int_0^a \frac{q}{f}\sin^2\theta\,dx.\end{aligned} \tag{3.3.2}$$

Thus the theorem will be proved if we show that, given any $\epsilon > 0$, we can choose $f > 0$ so that
$$\int_0^a |f - w^{1/2}|\,dx < \epsilon, \qquad \int_0^a \left|\frac{w}{f} - f\right|\,dx < \epsilon. \tag{3.3.3}$$

Let $w_1\ (>0)$ and w_2 denote the lower and upper bounds of $w^{1/2}$ in $[0, a]$, and let σ_n denote the n-th $(C, 1)$ partial sum of the Fourier series of $w^{1/2}$. Then $w_1 \le \sigma_n(x) \le w_2$. We now choose $f = \sigma_n$ where n is large enough to make
$$\int_0^a (w^{1/2} - \sigma_n)^2\,dx < \left(\frac{w_1\epsilon}{2w_2}\right)^2 \frac{1}{a}.$$

Then the first inequality in (3.3.3) certainly holds, and so does the second since
$$\left|\frac{w}{f} - f\right| = \frac{w^{1/2} + f}{f}|w^{1/2} - f| \le 2\frac{w_2}{w_1}|w^{1/2} - f|.$$

3.3. The coefficient w

Thus (3.3.2) gives
$$|2(m+1)\pi - \lambda^{1/2}I| < 3\epsilon\lambda^{1/2}$$
if $\lambda > \Lambda(\epsilon)$, and this proves (3.3.1). □

In the next theorem, we improve the $o(m)$ term in (3.3.1) by introducing differentiability of w.

Theorem 3.3.2. *In (3.3.2), let w be $AC[0,a]$. Then λ_{2m+1} and λ_{2m+2} both satisfy*
$$\sqrt{\lambda} = 2(m+1)\pi I^{-1} + o(1) \quad (m \to \infty). \tag{3.3.4}$$

Proof. Since w is differentiable now, we can choose $f = w^{1/2}$ in (3.2.5) and then (3.2.6) simplifies to
$$\theta' = \lambda^{1/2} w^{1/2} + \frac{w'}{4w} \sin 2\theta - \lambda^{-1/2} w^{-1/2} q \sin^2 \theta \tag{3.3.5}$$
which, on integration, gives
$$2(m+1)\pi = \lambda^{1/2} I + \frac{1}{4} \int_0^a \frac{w'}{w} \sin 2\theta \, dx - \lambda^{-1/2} \int_0^a w^{-1/2} q \sin^2 \theta \, dx.$$
The first integral on the right is $o(1)$ as $\lambda \to \infty$ by the Riemann-Lebesgue type Lemma 3.2.1 since (3.3.5) is covered by (3.2.12). Hence
$$2(m+1)\pi = \lambda^{1/2} I + o(1).$$
This proves (3.3.4). □

Before moving on to greater precision of the $o(1)$ term in (3.3.4), we give a result in which the smoothness of w falls in between what is considered in the above two theorems. In the next theorem, we take w to be piecewise smooth in the sense that w has a finite number N of discontinuities a_j in $[0,a)$, while w is absolutely continuous in each sub-interval $[a_j, a_{j+1}]$.

Theorem 3.3.3. *Let w be piecewise smooth with discontinuities a_j, where*
$$0 \leq a_1 < \cdots < a_N < a.$$
Let
$$s_j = \left(w(a_j + 0)/w(a_j - 0)\right)^{1/4} \tag{3.3.6}$$
and define α_j by
$$\alpha_j = \tan^{-1}\left(\tfrac{1}{2}|s_j - s_j^{-1}|\right) \in (0, \pi/2). \tag{3.3.7}$$
Then, for λ_{2m+1} and λ_{2m+2},
$$\left|\sqrt{\lambda} - 2(m+1)\pi I^{-1}\right| \leq I^{-1} \sum_{j=1}^N \alpha_j + o(1), \tag{3.3.8}$$
as $m \to \infty$. The same result holds for μ_{2m} and μ_{2m+1} but with $2(m+1)$ replaced by $2m+1$.

Proof. It is convenient to move the interval $[0, a]$ slightly (if necessary) to make $a_1 > 0$. Then $w(0) = w(a)$. This does not effect the λ_n and μ_n. We write $a_0 = 0$ and $a_{N+1} = a$.

We apply the Prüfer transformation (3.2.5) (with $f = w^{1/2}$) to (3.2.3) to obtain (3.3.5) again, but now with the proviso that (3.3.5) does not hold at the discontinuities a_j ($1 \leq j \leq N$). However, integration of (3.3.5) over (a_j, a_{j+1}) gives

$$\theta(a_{j+1} - 0, \lambda) - \theta(a_j + 0, \lambda) = \lambda^{1/2} \int_{a_j}^{a_{j+1}} w^{1/2} dx + o(1), \qquad (3.3.9)$$

on again using the Riemann-Lebesgue type Lemma 3.2.1.

The next step is to note that (3.2.11) continues to hold even though $\theta(x, \lambda)$ may have jumps at the a_j, provided that no such jump straddles a multiple of π. This can be effected because, by (3.2.5), $\theta(a_j + 0, \lambda)$ can be chosen to lie in the same quadrant as $\theta(a_j - 0, \lambda)$. Thus the method in section 2.4 for the proof of (3.2.11) continues to apply.

Summing over j in (3.3.9) and using (3.2.11), we obtain for λ_{2m+1} and λ_{2m+2}

$$2(m+1)\pi = \theta(a, \lambda) - \theta(0, \lambda)$$

$$= \lambda^{1/2} I + \sum_{j=1}^{N} (\theta(a_j + 0, \lambda) - \theta(a_j - 0, \lambda)) + o(1). \qquad (3.3.10)$$

To estimate the sum here, we write

$$t_\pm = \tan \theta(a_j \pm 0, \lambda), \quad w_\pm = w(a_j \pm 0)$$

for a particular j. By (3.2.5) (with $f = w^{1/2}$),

$$t_+/t_- = (w_+/w_-)^{1/2} = s_j^2$$

by (3.3.6), and so

$$\tan \left(\theta(a_j + 0, \lambda) - \theta(a_j - 0, \lambda) \right) = (t_+ - t_-)/(1 + t_+ t_-)$$

$$= \frac{(t_+ t_-)^{1/2}}{1 + t_+ t_-} (s_j - s_j^{-1}).$$

Since $2(t_+ t_-)^{1/2} \leq 1 + t_+ t_-$, we have

$$\left| \tan \left(\theta(a_j + 0, \lambda) - \theta(a_j - 0, \lambda) \right) \right| \leq \tan \alpha_j$$

by (3.3.7). Thus $|\theta(a_j + 0, \lambda) - \theta(a_j - 0, \lambda)| \leq \alpha_j$, and now (3.3.8) follows from (3.3.10). The proof for μ_{2m} and μ_{2m+1} is similar. $\square$

Moving on to greater differentiability, the next step would be to assume the existence of w'' in (3.2.3). But then w can be replaced by unity, as explained in the lead up to (3.2.4). Refinements of (3.3.4) then depend on the differentiability of q in (3.2.4), and this is taken up in the next section.

3.4 Titchmarsh's asymptotic formula

We begin by obtaining an asymptotic formula for λ_{2n+1} and λ_{2m+2} subject only to q in (3.2.4) being locally integrable. The formula is due to Titchmarsh and it involves the complex Fourier coefficients c_n defined by

$$c_n = \frac{1}{a}\int_0^a q(x)\exp(-2n\pi xi/a)\,dx \quad (-\infty < n < \infty). \qquad (3.4.1)$$

By shifting the value of λ, we can assume without loss of generality that q has mean-value zero, that is,

$$c_0 := \frac{1}{a}\int_0^a q(x)\,dx = 0. \qquad (3.4.2)$$

Theorem 3.4.1. *In (3.2.4), let q be locally integrable in $(0,a)$ and let (3.4.2) hold. Then, as $m \to \infty$, λ_{2m+1} and λ_{2m+2} satisfy*

$$\sqrt{\lambda} = 2(m+1)\pi/a \pm \tfrac{1}{4}a\pi^{-1}m^{-1}|c_{2m+2}| + O(m^{-2}). \qquad (3.4.3)$$

Proof. We use the Prüfer transformation (3.2.5) (with $f = w = 1$), so that (3.2.6) and (3.2.7) are now

$$\theta' = \lambda^{1/2} - \lambda^{-1/2}q\sin^2\theta, \qquad (3.4.4)$$

$$\rho'/\rho = \tfrac{1}{2}\lambda^{-1/2}q\sin 2\theta. \qquad (3.4.5)$$

We denote by α ($= \alpha(\lambda)$) the initial value of θ at $x = 0$:

$$\theta(0,\lambda) = \alpha. \qquad (3.4.6)$$

Then, for periodic y in (3.2.5), we require

$$\theta(a,\lambda) - \alpha = 2(m+1)\pi \qquad (3.4.7)$$

and

$$\rho(a,\lambda) = \rho(0,\lambda), \qquad (3.4.8)$$

the value $m + 1$ appearing as noted in (3.2.11) because the eigenfunctions y for both λ_{2m+1} and λ_{2m+2} have $2m+2$ zeros in $[0,a)$ (Theorem 2.4.2). We note that integration of (3.4.4) gives

$$\theta(x,\lambda) = \alpha + x\sqrt{\lambda} + O(\lambda^{-1/2}) \quad (0 \le x \le a). \qquad (3.4.9)$$

Next we write $\sin^2\theta = \tfrac{1}{2}(1 - \cos 2\theta)$ in (3.4.4) and integrate over $(0,a)$. By (3.4.2) and (3.4.7), this gives

$$2(m+1)\pi = a\sqrt{\lambda} + \tfrac{1}{2}\lambda^{-1/2}\int_0^a q(x)\cos 2\theta\,dx. \qquad (3.4.10)$$

A first consequence of this equation is that
$$a\sqrt{\lambda} = 2(m+1)\pi + O(m^{-1}). \tag{3.4.11}$$
Then, substituting (3.4.9) and this last equation for $\sqrt{\lambda}$ back into (3.4.10), we obtain
$$2(m+1)\pi = a\sqrt{\lambda} + \frac{a}{4\pi m}\int_0^a q(x)\cos(2\alpha + 2(m+1)\pi x/a)\,dx$$
$$+ O(m^{-2}). \tag{3.4.12}$$
Also, integration of (3.4.5) over $(0,a)$ and use of (3.4.8) similarly gives
$$\int_0^a q(x)\sin(2\alpha + 2(m+1)\pi x/a)\,dx = O(m^{-1}). \tag{3.4.13}$$

We now in effect eliminate α between (3.4.12) and (3.4.13) as follows. Let U and V denote the two integrals in (3.4.12) and (3.4.13), so that $U = \operatorname{Re} W$ and $V = \operatorname{Im} W$, where
$$W = ae^{-2i\alpha}c_{2m+2}$$
by (3.4.1). Then $V = O(m^{-1})$ and $-|V| \le |W| - |U| \le |V|$. Hence
$$|U| = |W| + O(m^{-1}) = a|c_{2m+2}| + O(m^{-1}).$$
Since $U = \pm|U|$, the theorem follows from (3.4.12). $\square$

We mention again that the corresponding formula for μ_{2m} and μ_{2m+1} has $2m+1$ and c_{2m+1} in place of $2(m+1)$ and c_{2m+2} in (3.4.3). We also note that (3.4.13) can give information on the value of α and hence on the initial values at $x = 0$ of the eigenfunctions corresponding to λ_{2m+1} and λ_{2m+2}. Let a_{2m+2} and b_{2m+2} be the Fourier sine and cosine coefficients of q, so that
$$c_{2m+2} = \tfrac{1}{2}(a_{2m+2} - ib_{2m+2}).$$
Then (3.4.13) gives
$$a_{2m+2}\sin 2\alpha + b_{2m+2}\cos 2\alpha = O(m^{-1}). \tag{3.4.14}$$
If, for example, $m|c_{2m+2}| \to \infty$ as $m \to \infty$, we can write (3.4.14) as
$$\sin(2\alpha + \beta_m) = O(1/m|c_{2m+2}|),$$
where $\tan\beta_m = b_{2m+2}/a_{2m+2}$. This gives two values for α which, to within this last O-term, are
$$\alpha_1 = -\tfrac{1}{2}\beta_m, \quad \alpha_2 = -\tfrac{1}{2}\beta_m + \tfrac{1}{2}\pi.$$
This, via (3.4.6) and (3.2.5), leads to two sets of initial values for the two eigenfunctions in question.

3.4. Titchmarsh's asymptotic formula

As an illustration relating to (3.4.3), we take q to be a two-valued step-function

$$q(x) = \begin{cases} c_1 & (0 \le x < a_1), \\ c_2 & (a_1 \le x < a), \end{cases} \qquad (3.4.15)$$

where

$$a_1 c_1 + (a - a_1) c_2 = 0 \qquad (3.4.16)$$

to satisfy (3.4.2). The discriminant $D(\lambda)$ is easily found to be

$$2\cos\left(a_1\sqrt{\lambda-c_1}\right)\cos\left((a-a_1)\sqrt{\lambda-c_2}\right)$$
$$-\left(\frac{\sqrt{\lambda-c_1}}{\sqrt{\lambda-c_2}} + \frac{\sqrt{\lambda-c_2}}{\sqrt{\lambda-c_1}}\right)\sin\left(a_1\sqrt{\lambda-c_1}\right)\sin\left((a-a_1)\sqrt{\lambda-c_2}\right).$$

Then the equation $D(\lambda) = 2$ for the periodic eigenvalues can, after some manipulation, be written as

$$\left(\sqrt{\lambda-c_1}+\sqrt{\lambda-c_2}\right)^2 \sin^2\left(\tfrac{1}{2}\left(a_1\sqrt{\lambda-c_1}+(a-a_1)\sqrt{\lambda-c_2}\right)\right)$$
$$= \left(\sqrt{\lambda-c_1}-\sqrt{\lambda-c_2}\right)^2 \sin^2\left(\tfrac{1}{2}\left(a_1\sqrt{\lambda-c_1}-(a-a_1)\sqrt{\lambda-c_2}\right)\right).$$

Thus

$$\sin\left(\tfrac{1}{2}\left(a_1\sqrt{\lambda-c_1}+(a-a_1)\sqrt{\lambda-c_2}\right)\right)$$
$$= \pm\frac{\sqrt{\lambda-c_1}-\sqrt{\lambda-c_2}}{\sqrt{\lambda-c_1}+\sqrt{\lambda-c_2}}\sin\left(\tfrac{1}{2}\left(a_1\sqrt{\lambda-c_1}-(a-a_1)\sqrt{\lambda-c_2}\right)\right).$$

Now $\sqrt{\lambda-c_1} = \sqrt{\lambda} - \frac{c_1}{2}\lambda^{-1/2} + O(\lambda^{-3/2})$, and similarly for $\sqrt{\lambda-c_2}$. Hence, by (3.4.16),

$$\sin\left(\tfrac{a}{2}\sqrt{\lambda}+O(\lambda^{-3/2})\right)$$
$$= \pm\frac{c_1-c_2}{4\lambda}(1+O(\lambda^{-1}))\sin\left(\tfrac{1}{2}(2a_1-a)\sqrt{\lambda}+O(\lambda^{-1/2})\right)$$
$$= \pm\frac{c_1-c_2}{4\lambda}\sin\left(\tfrac{1}{2}(2a_1-a)\sqrt{\lambda}\right)+O(\lambda^{-3/2}).$$

Then, writing $\sqrt{\lambda} = 2(m+1)\pi/a + \delta$, we find that λ_{2m+1} and λ_{2m+2} satisfy

$$\sin\tfrac{1}{2}a\delta = \pm\frac{(c_1-c_2)a^2}{16(m+1)^2\pi^2}\sin(2(m+1)\pi a_1/a) + O(m^{-3}).$$

Thus

$$\delta = \pm\frac{(c_1-c_2)a}{8(m+1)^2\pi^2}\sin(2(m+1)\pi a_1/a) + O(m^{-3}).$$

This agrees with (3.4.3) but we have a better O-term due to the special choice (3.4.15) of q here.

3.5 Differentiable q

When differentiability conditions are imposed on q, the estimate (3.4.3) for λ_{2m+1} and λ_{2m+2} can be improved, becoming increasingly accurate the more times that q is differentiable. The next theorem gives the general result of this nature.

Theorem 3.5.1. *Let q have an absolutely continuous $(r-1)$-th derivative in $(-\infty, \infty)$ and let (3.4.2) hold. Then there are numbers A_j independent of m and depending only on q such that, as $m \to \infty$, λ_{2m+1} and λ_{2m+2} satisfy*

$$\sqrt{\lambda} = 2(m+1)\pi/a + \sum_{j=2}^{r} A_j (2(m+1)\pi/a)^{-j-1} \pm \tfrac{a}{4}\pi^{-1} m^{-1} |c_{2m+2}| + O(m^{-r-2}). \quad (3.5.1)$$

Proof. Since q is differentiable now, the Prüfer transformation (3.2.8) becomes available. This one is more effective than (3.2.5) (with $f = w = 1$) for large λ because the second term in (3.2.9) is $O(\lambda^{-1})$ whereas the second term in (3.4.4) is only $O(\lambda^{-1/2})$. Integration of (3.2.9) and (3.2.10) over $(0, a)$, and substitution of (3.4.7) and (3.4.8) gives the two equations on which the proof of (3.5.1) depends:

$$2(m+1)\pi = \int_0^a (\lambda - q)^{1/2}\, dx - \frac{1}{4}\int_0^a \frac{q'}{\lambda - q} \sin 2\theta\, dx, \quad (3.5.2)$$

$$0 = \int_0^a \frac{q'}{\lambda - q} \cos 2\theta\, dx. \quad (3.5.3)$$

Beginning with (3.5.2), we use the familiar technique of integration by parts to determine the asymptotic form of the second integral on the right. To be systematic, we prove by induction on j $(2 \leq j \leq r)$ that

$$\int_0^a \frac{q'}{\lambda - q} \sin 2\theta\, dx = (-1)^{j-1} \int_0^a \frac{q^{(j)}}{(\lambda - q)^{(j+1)/2}} s^{(-j+1)}(\theta)\, dx$$

$$+ I_j + \sum_{k=3}^{5} \int_0^a R_{jk}(q,\theta)/(\lambda - q)^{(j+k)/2}\, dx, \quad (3.5.4)$$

where

(1) $s^{(-j+1)}(\theta)$ denotes the $(j-1)$th integral (with respect to θ) of $\sin 2\theta$ (and is therefore $\pm 2^{-(j-1)} \sin 2\theta$ or $\pm 2^{-(j-1)} \cos 2\theta$);

(2)
$$I_j = \sum_{\nu=5}^{j+5} \int_0^a P_\nu(q)/(\lambda - q)^{\nu/2}\, dx, \quad (3.5.5)$$

where $P_\nu(q)$ is a polynomial in $q', \ldots q^{(j-1)}$;

3.5. Differentiable q

(3) each $R_{jk}(q,\theta)$ ($3 \le k \le 5$) is a finite sum $\sum_n Q_{jn}(q) t_{jn}(\theta)$, where $Q_{jn}(q)$ is a polynomial in $q', \ldots, q^{(j-1)}$ and $t_{jn}(\theta)$ is a polynomial in $\sin 2\theta$ and $\cos 2\theta$ with mean-value zero with respect to the θ-interval $(0, 2\pi)$. The point about the mean-value being zero here is the consequence that the integral $t_{jn}^{(-1)}$ (w.r.t. θ) is also a trigonometric polynomial, and we choose the constant of integration so that the integral also has mean-value zero. We have already anticipated this convention in the definition of $s^{(-j+1)}(\theta)$ in (1) above.

The case $j = 1$ of (3.5.4) is of course trivial, and we start the induction argument with the first integral on the right in (3.5.4). By (3.2.9) and an integration by parts, this integral is

$$\int_0^a \frac{q^{(j)}}{(\lambda - q)^{(j+2)/2}} s^{(-j+1)}(\theta) \left(\theta' + \frac{1}{4} \frac{q'}{\lambda - q} \sin 2\theta \right) dx$$
$$= -\int_0^a \frac{q^{(j+1)}}{(\lambda - q)^{(j+2)/2}} s^{(-j)}(\theta) \, dx$$
$$+ \int_0^a \frac{q^{(j)} q'}{(\lambda - q)^{(j+4)/2}} \left(\tfrac{1}{4} s^{(-j+1)}(\theta) \sin 2\theta - \tfrac{1}{2}(j+2) s^{(-j)}(\theta) \right) dx,$$

the evaluated terms cancelling because of the periodicity. The first integral on the right here has the required form for $j + 1$. In the second integral, we subtract from $\tfrac{1}{4} s^{(-j+1)}(\theta) \sin 2\theta$ its mean value, and this subtracted term (if not zero) contributes to I_{j+1} in (3.5.5). The expression remaining in $(\ldots)$ contributes to $R_{j+1,3}$. Next (and similarly), we deal with a typical term $Q_{jn} t_{jn}$ in R_{j3} in (3.5.4). Thus

$$\int_0^a Q_{jn} t_{jn}(\theta)/(\lambda - q)^{(j+3)/2} \, dx = -\int_0^a \frac{Q'_{jn} t_{jn}^{(-1)}(\theta)}{(\lambda - q)^{(j+4)/2}} \, dx$$
$$+ \int_0^a \frac{Q_{jn} q'}{(\lambda - q)^{(j+6)/4}} \left(\tfrac{1}{4} t_{jn}(\theta) \sin 2\theta - \tfrac{1}{2}(j+4) t_{jn}^{(-1)}(\theta) \right) dx.$$

The two integrals here contribute, respectively, to $R_{j+1,3}$ and $R_{j+1,5}$, and the second integral may also contribute to I_{j+1}. Finally, the two integrals with $k = 4$ and $k = 5$ in (3.5.4) remain as they are and contribute to the $k = 3$ and $k = 4$ terms for $j+1$. This completes the induction proof of (3.5.4).

We now take $j = r$ in (3.5.4) and, to be definite, we suppose that $s^{(-r+1)}(\theta) = 2^{-(r-1)} \sin 2\theta$, i.e., $(r-1)/4$ is an integer; the following reasoning also covers the other cases. We also express (3.5.5) in terms of negative powers of λ to write (3.5.4) (with $j = r$) as

$$\int_0^a \frac{q'}{\lambda - q} \sin 2\theta \, dx = \sum_{\nu=5}^{r+1} B_\nu \lambda^{-\nu/2} + 2^{-(r-1)} \lambda^{-(r+1)/2} \int_0^a q^{(r)} \sin 2\theta \, dx$$
$$+ O(\lambda^{-(r+2)/2}), \quad (3.5.6)$$

where the B_ν are independent of λ and depend only on q.

The treatment of the integral in (3.5.3) is similar, and we obtain

$$\int_0^a \frac{q'}{\lambda - q} \cos 2\theta \, dx$$
$$= \sum_{\nu=6}^{r+1} C_\nu \lambda^{-\nu/2} + 2^{-(r-1)} \lambda^{-(r+1)/2} \int_0^a q^{(r)} \cos 2\theta \, dx + O(\lambda^{-(r+2)/2}), \quad (3.5.7)$$

where again the C_ν are independent of λ. There is however a new feature now in that we assert that all $C_\nu = 0$ as a result of the periodicity of q. A direct proof of this would require a more intricate inspection of the above integration by parts process. However, an indirect and much simpler proof observes that, by (3.5.3), the right-hand side of (3.5.7) is zero for a sequence of values of λ (viz. the periodic eigenvalues) tending to infinity. As a result, the dominant coefficient C_6 is zero and then similarly, in turn, all the C_ν are zero. There is just a gloss to add in the case of C_{r+1}, which is that the integral on the right in (3.5.7) is $o(1)$ ($\lambda \to \infty$) by Lemma 3.2.1. Then $\lambda \to \infty$ shows that $C_{r+1} = 0$ in its turn. Thus (3.5.3) and (3.5.7) reduce to

$$\int_0^a q^{(r)} \cos 2\theta \, dx = O(\lambda^{-1/2}). \quad (3.5.8)$$

The proof now moves to a conclusion in much the same way as the proof of Theorem 3.4.1. We again have (3.4.9) and (3.4.11), and then (3.5.2), (3.5.6) and (3.5.8) give

$$2(m+1)\pi = \int_0^a (\lambda - q)^{1/2} \, dx - \frac{1}{4} \sum_{\nu=5}^{r+1} B_\nu \lambda^{-\nu/2}$$
$$- 2^{-(r+1)} \lambda^{-(r+1)/2} \int_0^a q^{(r)} \sin(2\alpha + 2(m+1)\pi x/a) \, dx$$
$$+ O(\lambda^{-(r+2)/2}), \quad (3.5.9)$$

and

$$\int_0^a q^{(r)} \cos(2\alpha + 2(m+1)\pi x/a) \, dx = O(\lambda^{-1/2}).$$

These two equations correspond to (3.4.12) and (3.4.13), and the unknown α is eliminated as before. Also, the first integral on the right of (3.5.9) is expressed in terms of a binomial series involving negative powers of λ, except for the first term $a\sqrt{\lambda}$. The resulting modification of (3.5.9) is then "inverted" to express the negative powers in terms of $2(m+1)\pi/a$ rather than $\sqrt{\lambda}$. This completes the proof of (3.5.1) once we note that the Fourier coefficient $c_{2m+2}^{(r)}$ of $q^{(r)}$ is $(4(m+1)\pi i/a)^r c_{2m+2}$. □

3.6. Length of the instability intervals

The first few of the A_j in (3.5.1) can be determined by examining the early integrations by parts in (3.5.2). Thus

$$A_1 = \frac{1}{2a}\int_0^a q(x)\,dx, \qquad A_2 = 0, \qquad A_3 = \frac{1}{8a}\int_0^a q^2(x)\,dx - A_1^2 \qquad (3.5.10)$$

and, in particular, $A_1 = 0$ if (3.4.2) is assumed.

3.6 Length of the instability intervals

The length l_n of the n-th instability interval I_n is given by

$$l_n = \begin{cases} \lambda_{2m+2} - \lambda_{2m+1} \\ \mu_{2m+1} - \mu_{2m} \end{cases}$$

according as $n = 2m+2$ or $n = 2m+1$ $(m \geq 0)$, and the asymptotic results of the previous sections can be put together to give the following estimates for l_n $(n \to \infty)$ in terms of the equation (3.2.3).

(i) $l_n = o(n^2)$ (see Theorem 3.3.1).

(ii) Let w be piecewise smooth as in Theorem 3.3.3. Then

$$l_n \leq 4n\pi I^{-2} \sum_{j=1}^{N} \alpha_j.$$

(iii) Let w be $AC[0, a]$ as in Theorem 3.3.2. Then

$$l_n = o(n).$$

(iv) Let $w = 1$ and let $q^{(r-1)}$ be $AC[0, a]$ for some $r \geq 0$. Then, by Theorems 3.4.1 and 3.5.1,

$$l_n = 2c_n + O(n^{-r}).$$

It is natural to ask whether (iv) can be further refined to give an exponentially small estimate for l_n if q is analytic in some sense. In the case of the Mathieu equation, where $q(x) = (const.)\cos 2x$, a precise estimate of this kind does exist and is given in the next section.

For general coefficients q which are infinitely differentiable, there is also an exponentially small estimate for l_n, but the estimate is not claimed to be precise. The estimate is given in the next theorem, where the condition (3.6.1) on the derivatives covers suitable analytic coefficients.

Theorem 3.6.1. Let $w = 1$ and let q be infinitely differentiable with
$$|q^{(r)}(x)| \leq K r! \, d^{-r} \tag{3.6.1}$$
for all x and $r \geq 0$, where K and d are independent of x and r. Then there are positive constants A and B such that
$$l_n \leq A \exp(-Bn) \quad (n \to \infty). \tag{3.6.2}$$

The proof is given in several stages in the rest of this section. For the basis of the proof, we refer forward to Corollary 4.7.3, where the length $l(\mu)$ of an instability interval with mid-point μ is shown to satisfy
$$l(\mu) \leq 2 \liminf_{m \to \infty} \left\| \left(-\frac{d^2}{dx^2} + q - \mu \right) f_m \right\|, \tag{3.6.3}$$
where $\{f_m\}$ is any sequence of functions with compact support such that f'_m is $AC[0, \infty)$, and
$$\|f_m\| = 1. \tag{3.6.4}$$
The norm here is that of the Hilbert space $L^2(0, \infty)$.

The first stage is to make the choice of f_m and outline how it is used. Then in a lemma we estimate certain terms which are involved in, and justify, the choice of f_m. Finally, we estimate two error terms to obtain a theorem on $l(\mu)$, of which Theorem 3.6.1 is a consequence.

We choose f_m in (3.6.3) to have the form
$$f_m(x) = \alpha_m P(x, \mu) \, h_m(x) \exp\left(i \int_0^x Q(t, \mu) \, dt \right), \tag{3.6.5}$$
where $P(x, \mu)$ and $Q(x, \mu)$ are real-valued, α_m is the normalisation constant making $\|f_m\| = 1$ as in (3.6.4), and
$$h_m(x) = \begin{cases} (1 - (x - 2m)^2/m^2)^3 & (|x - 2m| \leq m), \\ 0 & (|x - 2m| > m). \end{cases} \tag{3.6.6}$$
Thus f_m has compact support $[m, 3m]$ with sufficient smoothness. The functions $P(x, \mu)$ and $Q(x, \mu)$ are independent of m, $Q(x, \mu)$ has the form given by
$$Q^2(x, \mu) = \mu \left(1 + \sum_{j=1}^N a_j(x) \mu^{-j} \right), \tag{3.6.7}$$
where the $a_j(x)$ have period a and are to be chosen in terms of q, N is to be chosen in terms of μ and, finally, P is defined by
$$P^4 Q^2 = \mu. \tag{3.6.8}$$

3.6. Length of the instability intervals

We note that (3.6.8) implies that

$$2\frac{P'}{P} + \frac{Q'}{Q} = 0. \tag{3.6.9}$$

It also follows from (3.6.4)-(3.6.8) that

$$\alpha_m = O(m^{-1/2}) \quad (m \to \infty) \tag{3.6.10}$$

and from (3.6.6) that

$$h'_m(x) = O(m^{-1}), \quad h''_m(x) = O(m^{-2}) \tag{3.6.11}$$

uniformly in x as $m \to \infty$.

Considering now the choice (3.6.5) in the right-hand side of (3.6.3) and using (3.6.9), we have

$$\left(-\frac{d^2}{dx^2} + q - \mu\right) f_m = \left(Q^2 - \mu + q - \frac{P''}{P}\right) f_m + H_m, \tag{3.6.12}$$

where

$$|H_m| = \alpha_m |2(P' - iPQ)h'_m - Ph''_m|.$$

By (3.6.10) and (3.6.11), we have $\|H_m\| = O(m^{-1})$ $(m \to \infty)$, and hence (3.6.3), (3.6.4) and (3.6.12) give

$$l(\mu) \le 2 \sup_{0 \le x < \infty} \left|Q^2 - \mu + q - \frac{P''}{P}\right|. \tag{3.6.13}$$

On the basis of (3.6.7) and (3.6.8), we need the expansion of P''/P in powers of μ^{-1}. Thus

$$\frac{P''}{P} = -\frac{1}{4}\left(\sum_{j=1}^{N} a''_j \mu^{-j}\right)\bigg/\left(1 + \sum_{j=1}^{N} a_j \mu^{-j}\right)$$

$$+ \frac{5}{16}\left(\sum_{j=1}^{N} a'_j \mu^{-j}\right)^2 \bigg/ \left(1 + \sum_{j=1}^{N} a_j \mu^{-j}\right)^2 \tag{3.6.14}$$

$$= -\frac{1}{4}\left(\sum_{j=1}^{N-1} b_j \mu^{-j} + \mu^{-N} E_N\right) + \frac{5}{16}\left(\sum_{j=2}^{N-1} c_j \mu^{-j} + \mu^{-N} F_N\right) \tag{3.6.15}$$

where, in the remainder terms, the sizes of E_N and F_N have to be estimated. We note that the b_j and c_j involve only $a_1, \ldots, a_j$ (and their derivatives). The next step then in this outline is to choose the a_j so that there are no powers μ^{-j} $(1 \le j \le N-1)$ appearing in (3.6.13). Thus, by (3.6.7) and (3.6.15),

$$a_1 = -q, \quad a_2 = -\frac{1}{4}b_1 = -\frac{1}{4}a''_1 = \frac{1}{4}q'' \tag{3.6.16}$$

and, for $1 \leq j \leq N-1$,
$$a_{j+1} = -\frac{1}{4} b_j + \frac{5}{16} c_j \tag{3.6.17}$$
where $c_1 = 0$ and $c_2 = q'^2$. We have here a recursive definition of the a_j in terms of q and its derivatives, and the result is that (3.6.13) becomes
$$l(\mu) \leq \left(\frac{1}{2} \sup_x |E_N(x)| + \frac{5}{8} \sup_x |F_N(x)| \right) \mu^{-N}. \tag{3.6.18}$$

We have to justify this recursive procedure and then obtain the requisite estimates on $|E_N|$ and $|F_N|$.

Finally in these introductory remarks, we introduce an expansion similar to the first one in (3.6.15):
$$\left(\sum_{j=1}^{N} a'_j \mu^{-j} \right) \bigg/ \left(1 + \sum_{j=1}^{N} a_j \mu^{-j} \right) = \sum_{j=1}^{N-1} d_j \mu^{-j} + \mu^{-N} G_N. \tag{3.6.19}$$

The second series in (3.6.15) is obtained by squaring this last one. Now we can move on to the lemma already referred to, and which gives estimates for the introduced coefficients and their derivatives.

Lemma 3.6.2. *Let* (3.6.1) *hold with* $d \leq 1$ *and define*
$$\delta = \max\left(2K, \frac{9}{8} \right). \tag{3.6.20}$$
Then, for all $r \geq 0$ *and* $1 \leq j \leq N$,

(i)
$$|a_j^{(r)}(x)| \leq \frac{1}{2} \delta^j (2j-2+r)! \, d^{-(2j-2+r)} \tag{3.6.21}$$

(ii)
$$|b_{j-1}^{(r)}(x)| \leq \delta^{j-1} (2j-2+r)! \, d^{-(2j-2+r)} \tag{3.6.22}$$

(iii)
$$|d_{j-1}^{(r)}(x)| \leq \delta^{j-1} (2j-3r)! \, d^{-(2j-3+r)} \tag{3.6.23}$$

where $b_0 := 0$ *and* $d_0 := 0$.

Proof. We note that the assumption $d \leq 1$ is not a restriction on q. We argue by induction on j. For $j=1$ and $j=2$, we have from (3.6.16) and (3.6.19)
$$a_1 = -q, \qquad a_2 = \frac{1}{4} q'', \qquad b_1 = -q'', \qquad d_1 = -q',$$
showing that the lemma holds for $j=1$ and $j=2$. Now we assume that the lemma holds up to a general j and prove it for $j+1$.

3.6. Length of the instability intervals

We deal first with b_j. If, in the quotient in (3.6.14) which defines the b_j, we multiply by the denominator and equate coefficients of μ^{-j}, we obtain

$$b_j = -\sum_{k=1}^{j-1} b_{j-k} a_k + a_j''.$$

Hence

$$b_j^{(r)} = -\sum_{k=1}^{j-1}\sum_{s=0}^{r} \binom{r}{s} b_{j-k}^{(s)} a_k^{(r-s)} + a_j^{(r+2)}$$

and so, by hypothesis and the fact that $d \leq 1$,

$$d^{2j+r} \frac{\delta^{-j}|b_j^{(r)}|}{(2j+r)!} \leq \frac{r!}{2(2j+r)!}\sum_{k=1}^{j-1}\sum_{s=0}^{r}\frac{(2j-2k+s)!(2k-2+r-s)!}{s!(r-s)!} + \frac{1}{2} \quad (3.6.24)$$

$$= \frac{1}{2}\sum_{k=1}^{j-1}\sum_{s=0}^{r}\frac{(s+1)(\ldots)(s+2j-2k)}{(r+1)(\ldots)(r+2j-2k)}\cdot\frac{(r-s+1)(\ldots)(r-s+2k-2)}{(r+2j-2k+1)(\ldots)(r+2j)} + \frac{1}{2}$$

$$< \frac{1}{2}\sum_{k=1}^{j-1}\sum_{s=0}^{r}\frac{1}{(r+2j-1)(r+2j)} + \frac{1}{2}$$

$$= \frac{1}{2}\frac{(j-1)(r+1)}{(r+2j-1)(r+2j)} + \frac{1}{2} < 1 \quad (3.6.25)$$

as required for (3.6.22).

There is a similar proof for $d_j^{(r)}$ in (3.6.23), and we omit these details. However, since c_j appears in (3.6.17), we have to consider the relation connecting the c_j and d_j. Thus, by (3.6.14), (3.6.15) and (3.6.19), we have

$$c_j = \sum_{k=1}^{j-1} d_{j-k} d_k \quad (2 \leq j \leq N-1)$$

and hence, by (3.6.23),

$$d^{2j-2+r}\frac{\delta^{-j}|c_j^{(r)}|}{(2j+r)!} \leq \frac{r!}{(2j+r)!}\sum_{k=1}^{j-1}\sum_{s=0}^{r}\frac{(2j-2k-1+s)!\,(2k-1+r-s)!}{s!\,(r-s)!} < 1$$

as in (3.6.24) and (3.6.25). Thus $|c_j^{(r)}| \leq \delta^j (2j+r)!\, d^{-(2j-2+r)}$.

We use the last estimate, together with (3.6.25), in (3.6.17). This gives

$$\left|a_{j+1}^{(r)}\right| \leq \frac{9}{16}\delta^j(2j+r)!\,d^{-(2j+r)} \leq \frac{1}{2}\delta^{j+1}(2j+r)!\,d^{-(2j+r)}$$

by (3.6.20). Thus, finally, (3.6.21) holds for $j+1$, completing the induction argument and establishing the lemma. □

Theorem 3.6.3. Let (3.6.1) hold with $d \leq 1$ and let δ be as in (3.6.20). Then, if $\mu > 4\delta N^2 d^{-2}$, there is a constant C such that

$$l(\mu) \leq C(1 - 4\delta N^2/d^2\mu)^{-1} N^{-1} (4\delta N^2/d^2\mu)^N. \tag{3.6.26}$$

Proof. We have to estimate E_N and F_N in (3.6.18). Now E_N and F_N arise in (3.6.15) and, dealing with E_N first, we replace μ^{-1} by a complex variable z in the ratio which defines E_N, that is, we consider

$$f(z) := \Big(\sum_{j=1}^{N} a_j'' z^j\Big) \Big/ \Big(1 + \sum_{j=1}^{N} a_j z^j\Big) \qquad (|z| \leq R) \tag{3.6.27}$$

and R is chosen so that f has no singularities. By (3.6.21),

$$\Big|\sum_{j=1}^{N} a_j z^j\Big| \leq \frac{1}{2} \sum_{j=1}^{N} \delta^j (2j-2)! \, d^{-(2j-2)} R^j.$$

The terms in the series on the right are decreasing if R satisfies

$$(2N-2)(2N-3)\delta d^{-2} R \leq 1,$$

and this is certainly achieved by choosing

$$R = (4N^2 \delta)^{-1} d^2. \tag{3.6.28}$$

Then

$$\Big|\sum_{j=1}^{N} a_j z^j\Big| < \frac{1}{2}\delta RN = \frac{1}{8} N^{-1} d^2 \leq \frac{1}{8}. \tag{3.6.29}$$

Thus, in (3.6.27), f has no singularities when R is as in (3.6.28).

Again, for the numerator in f,

$$\Big|\sum_{j=1}^{N} a_j'' z^j\Big| \leq \frac{1}{2} \sum_{j=1}^{N} \delta^j (2j)! \, d^{-2j} R^j \tag{3.6.30}$$

by (3.6.21). The terms are decreasing if

$$2N(2N-1)\delta d^{-2} R \leq 1,$$

and this is also satisfied by (3.6.28). Then (3.6.30) is bounded above by $\delta R d^{-2} N = \frac{1}{4} N^{-1}$. This together with (3.6.29) gives

$$|f(z)| \leq \frac{2}{7} N^{-1} \tag{3.6.31}$$

3.6. Length of the instability intervals

when $|z| \leq (4N^2\delta)^{-1}d^2$.

Similarly, in connection with F_N in (3.6.14) and (3.6.15), we define

$$g(z) := \Big(\sum_{j=1}^{N} a'_j z^j\Big)^2 \Big/ \Big(1 + \sum_{j=1}^{N} a_j z^j\Big)^2$$

and, corresponding to (3.6.31),

$$|g(z)| \leq \frac{1}{49} N^{-2} d^2. \tag{3.6.32}$$

We can now move on to E_N itself. Let $\mu > 4N^2\delta d^{-2}$ ($= R^{-1}$) and write $t = \mu^{-1}$. Then the point t lies within the circle $|z| = R$, and Cauchy's theorem gives

$$f(t) = \frac{1}{2\pi i} \int_{|z|=R} \frac{f(z)}{z-t} \, dz$$

$$= f(0) + t f'(0) + \cdots + \frac{t^{N-1}}{(N-1)!} f^{(N-1)}(0) + \frac{1}{2\pi i} \int_{|z|=R} \frac{(t/z)^N f(z)}{z-t} \, dt$$

after expressing $(z-t)^{-1}$ as a geometric series. On comparing with (3.6.15), we see that

$$E_N = \frac{1}{2\pi i} \int_{|z|=R} \frac{z^{-N} f(z)}{z-t} \, dt.$$

Hence, by (3.6.31),

$$|E_N| \leq \frac{2}{7} R^{-N+1} N^{-1} (R-t)^{-1}.$$

Similarly, from (3.6.15) and (3.6.32),

$$|F_N| \leq \frac{1}{49} R^{-N+1} N^{-2} d^2 (R-t)^{-1}.$$

Then, since $N \geq 1$ and $d \leq 1$, (3.6.18) gives

$$l(\mu) \leq \frac{61}{392} R^{-N+1} N^{-1} (R-t)^{-1} \mu^{-N},$$

and now (3.6.26) follows from (3.6.28) since $t = \mu^{-1}$. $\square$

Finally, we are in a position to deduce (3.6.2) from (3.6.26). In the case of even $n = 2m+2$ (odd being similar), we have $l(\mu) = l_{2m+2}$ and $\mu = \frac{1}{2}(\lambda_{2m+2} + \lambda_{2m+1})$. We note that, by (3.5.1), (3.4.2) and (3.5.10),

$$\mu > 4(m+1)^2 \pi^2 / a^2 \tag{3.6.33}$$

for non-trivial q when m is sufficiently large. In (3.6.26) we choose

$$N = (m+1)\pi d/(a\sqrt{2\delta}).$$

Then

$$N^2 \le (m+1)^2 \pi^2 d^2/(a^2 2\delta) < d^2\mu/8\delta$$

by (3.6.33), and (3.6.26) gives

$$l_{2m+2} < 2CN^{-1}\exp(-N\log 2) < A\exp(-Bn),$$

as required.

The obvious situation where (3.6.1) is satisfied is when $q(x)$ is the restriction of a $q(z)$ which is analytic in the strip $\{z \in \mathbb{C} \mid d \le \operatorname{Im} z \le d\}$. Then Cauchy's theorem gives

$$\left|q^{(r)}(x)\right| = \frac{r!}{2\pi}\left|\int_{|z-x|=d}\frac{q(z)}{(z-x)^{r+1}}\,dz\right| \le r!\left(\max_z |q(z)|\right) d^{-r}.$$

Indeed, (3.6.1) is equivalent to $q(x)$ having an analytic continuation $q(z)$ defined via Taylor series.

3.7 The Mathieu equation

In this section we give a precise asymptotic expression for the length l_n of the n-th instability interval in the case of the Mathieu equation

$$y''(x) + (\lambda - 2c\cos 2x)\,y(x) = 0 \qquad (3.7.1)$$

with period π. In section 1.7, we saw that $l_n > 0$ for all n.

Theorem 3.7.1. *As $n \to \infty$,*

$$l_n = 8\left(\frac{|c|}{4}\right)^n \frac{1}{((n-1)!)^2}(1+O(n^{-2})). \qquad (3.7.2)$$

Before starting the proof, we note that Stirling's formula for $(n-1)!$ gives

$$l_n = \frac{4n}{\pi}\left(\exp\left(-2n\log n + 2n + n\log\frac{|c|}{4}\right)\right)\left(1 - \frac{1}{6n} + O(n^{-2})\right),$$

showing that l_n is even more than exponentially small as $n \to \infty$.

Proof. We consider even $n = 2m$:

$$l_{2m} = \lambda_{2m} - \lambda_{2m-1}.$$

3.7. The Mathieu equation

The proof for odd $n = 2m+1$ is similar. By Theorem 2.8.1, there are even and odd periodic eigenfunctions ψ_m^{I} and ψ_m^{II} associated with λ_{2m-1} and λ_{2m} (not necessarily in that order). The order here is not important, and we simply write λ' and λ'' for the eigenvalues corresponding respectively to ψ_m^{I} and ψ_m^{II}. Thus $l_{2m} = |\lambda'' - \lambda'|$.

We substitute the Fourier expansions

$$\psi_m^{\text{I}}(x) = \frac{1}{2}a_0 + \sum_{r=1}^{\infty} a_r \cos(2rx),$$

$$\psi_m^{\text{II}}(x) = \sum_{r=1}^{\infty} b_r \sin(2rx)$$

into (3.7.1) with $\lambda = \lambda'$ and $\lambda = \lambda''$ respectively. As with a similar situation in section 1.7, we obtain

$$\frac{1}{2}\lambda' a_0 - c a_1 = 0 \tag{3.7.3}$$

and, for $r \geq 1$,

$$(4r^2 - \lambda') a_r + c(a_{r-1} + a_{r+1}) = 0, \tag{3.7.4}$$

$$(4r^2 - \lambda'') b_r + c(b_{r-1} + b_{r+1}) = 0 \tag{3.7.5}$$

where b_0 is defined to be zero. Now eliminate $4r^2$:

$$(\lambda' - \lambda'') a_r b_r = c\Big((a_{r-1} + a_{r+1})b_r - (b_{r-1} + b_{r+1})a_r\Big)$$
$$= c\Big((a_{r-1}b_r - a_r b_{r-1}) - (a_r b_{r+1} - a_{r+1} b_r)\Big).$$

Hence

$$(\lambda' - \lambda'') \sum_{r=0}^{\infty} a_r b_r = c(a_0 b_1 - a_1 b_0) = c a_0 b_1. \tag{3.7.6}$$

So far, this is not dissimilar to what we did in section 1.7. Now however, we have to determine the asymptotic forms of the sum on the left of (3.7.6) as well as a_0 and b_1. We proceed with the proof in a number of steps.

Step 1. If ψ_m^{I} and ψ_m^{II} are normalized so that

$$\psi_m^{\text{I}}(0) = 1, \quad \psi_m^{\text{II}'}(0) = 1, \tag{3.7.7}$$

then we show that

$$\sum_{r=0}^{\infty} a_r b_r = 1 + O(m^{-2}). \tag{3.7.8}$$

We first note that, by Theorem 3.5.1, λ' and λ'' both satisfy the simple asymptotic equation

$$\sqrt{\lambda} = 2m + O(m^{-3}). \tag{3.7.9}$$

Then, referring forward to Theorem 3.8.2 applied to ψ_m^I and ψ_m^{II}, and using (3.7.9), we obtain

$$\psi_m^I(x) = \cos 2mx - \frac{1}{4}km^{-1} \sin 2x \sin 2mx + O(m^{-2}), \qquad (3.7.10)$$

$$\psi_m^{II}(x) = \sin 2mx + \frac{1}{4}km^{-1} \sin 2x \cos 2mx + O(m^{-2}).$$

It follows that

$$\left.\begin{array}{l} a_m := \frac{2}{\pi}\int_0^\pi \psi_m^I(x) \cos(2mx)\, dx = 1 + O(m^{-2}), \\[4pt] b_m := \frac{2}{\pi}\int_0^\pi \psi_m^{II}(x) \sin(2mx)\, dx = 1 + O(m^{-2}). \end{array}\right\} \qquad (3.7.11)$$

Further, the Parseval formula for ψ_m^I gives

$$\frac{1}{2}a_0^2 + \sum_1^\infty a_r^2 = \frac{2}{\pi}\int_0^\pi \psi_m^{I\,2}(x)\, dx = 1 + O(m^{-2})$$

by (3.7.10). Hence, by (3.7.11),

$$\frac{1}{2}a_0^2 + \sum_{r\neq m} a_r^2 = O(m^{-2}). \qquad (3.7.12)$$

Similarly, from ψ_m^{II},

$$\sum_{r\neq m} b_r^2 = O(m^{-2}). \qquad (3.7.13)$$

Now (3.7.8) follows from (3.7.11)-(3.7.13) and the Cauchy inequality.

So far, we have dealt with the left-hand side of (3.7.6). To estimate a_0 and b_1, on the right-hand side, we use the recurrence formulae (3.7.4) and (3.7.5)– in effect starting with (3.7.11) and working back to a_0 and b_1. The next step in the proof of the theorem gives the general form of a_r which we require. The form of b_r is similar.

Step 2. We prove by induction on r that, for $r \leq m$,

$$a_r = \frac{1}{2} a_0 c^{-r} \lambda'(\lambda' - 4)(\ldots)(\lambda' - 4(r-1)^2)(1 + E_r), \qquad (3.7.14)$$

where

$$|E_r| \leq c^2\left(1 + \frac{rc^2}{\lambda'}\right) S_r \qquad (3.7.15)$$

and

$$S_r = \frac{1}{\lambda'^2} + \frac{2}{(\lambda' - 4)^2} + \cdots + \frac{2}{(\lambda' - 4(r-2)^2)^2} + \frac{1}{(\lambda' - 4(r-1)^2)^2}. \qquad (3.7.16)$$

3.7. The Mathieu equation

First, (3.7.14) holds trivially for $r = 1$ since
$$a_1 = \frac{1}{2} a_0 c^{-1} \lambda'$$
by (3.7.3). Also, for $r = 2$, we have from (3.7.4)
$$c_2 = -a_0 + c^{-1}(\lambda' - 4) a_1 = -a_0 + \frac{1}{2} a_0 c^{-2} \lambda'(\lambda' - 4)$$
$$= \frac{1}{2} a_0 c^{-2} \lambda'(\lambda' - 4)\left(1 - \frac{2c^2}{\lambda'(\lambda' - 4)}\right),$$
and hence
$$|E_2| \leq c^2 \left(\frac{1}{\lambda'^2} + \frac{1}{(\lambda' - 4)^2}\right),$$
certainly satisfying (3.7.15).

Now we assume that (3.7.14)-(3.7.16) hold for a_{r-1} and a_r and deduce the result for a_{r+1} ($r \leq m - 1$). From (3.7.4) we have
$$a_{r+1} = -a_{r-1} + c^{-1}(\lambda' - 4r^2) a_r$$
$$= -\frac{1}{2} a_0 c^{-(r-1)} \lambda'(\ldots)(\lambda' - 4(r-2)^2)(1 + E_{r-1})$$
$$+ c^{-1}(\lambda' - 4r^2) \frac{1}{2} a_0 c^{-r} \lambda'(\ldots)(\lambda' - 4(r-1)^2)(1 + E_r)$$
$$= \frac{1}{2} a_0 c^{-(r+1)} \lambda'(\ldots)(\lambda' - 4r^2)(1 + E_{r+1}),$$
where
$$E_{r+1} = E_r - \frac{c^2}{(\lambda' - 4(r-1)^2)(\lambda' - 4r^2)} (1 + E_{r-1}). \tag{3.7.17}$$
We have to show that E_{r+1} satisfies (3.7.15) with $r + 1$ in place of r.

By (3.7.9) and the fact that $r \leq m - 1$, it follows from (3.7.15) (with $r - 1$) that
$$|E_{r-1}| \leq 2c^2 S_{r-1} < 2c^2 S_r$$
when n is large enough. Also
$$(\lambda' - 4(r-1)^2)(\lambda' - 4r^2)$$
$$= \left(\sqrt{\lambda'} - 2(r-1)\right)\left(\sqrt{\lambda'} + 2(r-1)\right)\left(\sqrt{\lambda'} - 2r\right)\left(\sqrt{\lambda'} + 2r\right)$$
$$> 3\lambda'.$$
Hence (3.7.17) gives
$$|E_{r+1}| \leq c^2 \left(1 + \frac{rc^2}{\lambda'}\right) S_r + \frac{c^2}{(\lambda' - 4(r-1)^2)^2} + \frac{c^2}{(\lambda' - 4r^2)^2} + \frac{2c^2}{3\lambda'} S_r$$
$$< c^2 \left(1 + \frac{rc^2}{\lambda'}\right) S_{r+1} + \frac{2c^2}{3\lambda'} S_{r+1} < c^2 \left(1 + \frac{(r+1)c^2}{\lambda'}\right) S_{r+1}$$

by (3.7.16) (with $r+1$). This completes the induction proof of (3.7.14)-(3.7.16).

Step 3. The next step is to show that

$$E_m = O(m^{-2}) \qquad (m \to \infty) \qquad (3.7.18)$$

and to substitute what we have found so far into (3.7.6). In the following, we use the fact that, by (3.7.9), $\lambda' > (2m-1)^2$ for m large enough. Then, by (3.7.15) and (3.7.16) again for m large enough, we have

$$|E_m| < 4c^2 \sum_{s=0}^{m-1} \frac{1}{(\lambda' - 4s^2)^2} < 4c^2 \sum_{s=0}^{m-1} \frac{1}{\left((2m-1)^2 - 4s^2\right)^2}$$

$$= 4c^2 \sum_{s=0}^{m-1} \frac{1}{(2m-1-2s)^2(2m-1+2s)^2}$$

$$< 4c^2 \frac{1}{(2m-1)^2} \sum_{s=0}^{m-1} \frac{1}{(2m-1-2s)^2} < 4c^2 \frac{1}{(2m-1)^2} \frac{\pi^2}{6},$$

and this proves (3.7.18).

It now follows from (3.7.14), (3.7.11) and (3.7.18) that

$$a_0 = 2c^m \left(\lambda'(\lambda' - 4)(\ldots)(\lambda' - 4(m-1)^2)\right)^{-1} (1 + O(m^{-2})).$$

The similar result for b_1 is

$$b_1 = c^{m-1} \left((\lambda'' - 4)(\ldots)(\lambda'' - 4(m-1)^2)\right)^{-1} (1 + O(m^{-2})).$$

Hence, by (3.7.8) also, (3.7.6) gives

$$\lambda' - \lambda'' = 2c^{2m} \Big(\lambda'(\lambda' - 4)(\lambda'' - 4)(\ldots)$$
$$\times (\lambda' - 4(m-1)^2)(\lambda'' - 4(m-1)^2)\Big)^{-1} (1 + O(m^{-2})). \qquad (3.7.19)$$

Step 4. The final step is to use the asymptotic form (3.7.9) in (3.7.19) to arrive at (3.7.2). By (3.7.9), both λ' and λ'' satisfy $\lambda = 4m^2 + O(m^{-2})$. Hence, for $r \leq m-1$,

$$\lambda - 4r^2 = 4(m^2 - r^2)(1 + O(m^{-3})).$$

Then (3.7.19) becomes

$$\lambda' - \lambda'' = 2c^{2m} 4^{-2m+1} \Big(m(m^2 - 1)(\ldots)(m^2 - (m-1)^2)\Big)^{-2} (1 + O(m^{-1}))$$

$$= 8\left(\frac{c}{4}\right)^{2m} \Big(m(m-1)(m+1)(\ldots)(1)(2m-1)\Big)^{-2} (1 + O(m^{-2}))$$

$$= 8\left(\frac{c}{4}\right)^{2m} \frac{1}{((2m-1)!)^2} (1 + O(m^{-2})).$$

This proves (3.7.2) for even $n = 2m$ and, as mentioned at the outset, the proof for odd $n = 2m + 1$ is similar. □

The rapidly decreasing nature of l_n, shown asymptotically by (3.7.2), is already illustrated numerically by the example $c = 1$ and the low value $n = 10$. Here $\lambda_9 = 25.0208408$ and $\lambda_{10} = 25.0208543$, giving $\lambda_{10} = 13 \times 10^{-6}$.

3.8 Asymptotic formulae for solutions

In this section we again take $w = 1$, so that the differential equation is (3.2.4), and we obtain asymptotic formulae for the basic solutions $\phi_1(x, \lambda)$ and $\phi_2(x, \lambda)$ as $\lambda \to \infty$. As for λ_n and μ_n, the formulae become more precise as more differentiability is imposed on q. We begin however without the imposition of any such conditions. The integral equation formulations for ϕ_1 and ϕ_2, derived from the variation of constants formula, are

$$\phi_1(x, \lambda) = \cos x\sqrt{\lambda} + \lambda^{-1/2} \int_0^x \sin\left((x-t)\sqrt{\lambda}\right) q(t)\phi_1(t, \lambda)\, dt, \qquad (3.8.1)$$

$$\phi_2(x, \lambda) = \lambda^{-1/2} \sin x\sqrt{\lambda} + \lambda^{-1/2} \int_0^x \sin\left((x-t)\sqrt{\lambda}\right) q(t)\phi_2(t, \lambda)\, dt. \qquad (3.8.2)$$

Let $M_1(\lambda)$ denote the greatest value of $|\phi_1(x, \lambda)|$ for x in $[0, a]$. Then, by (3.8.1),

$$M_1(\lambda) \leq 1 + \lambda^{-1/2} M_1(\lambda) \int_0^a |q(t)|\, dt$$

and hence

$$M_1(\lambda) \leq \left(1 - \lambda^{-1/2} \int_0^a |q(t)|\, dt\right)^{-1} \leq 2 \qquad (3.8.3)$$

if, say, $\lambda^{1/2} \geq 2 \int_0^a |q(t)|\, dt$. Then (3.8.1) gives

$$\phi_1(x, \lambda) = \cos x\sqrt{\lambda} + O(\lambda^{-1/2}). \qquad (3.8.4)$$

Here and subsequently, O and o terms are uniform with respect to x in $[0, a]$. Similarly, from (3.8.2),

$$\phi_2(x, \lambda) = \lambda^{-1/2} \sin x\sqrt{\lambda} + O(\lambda^{-1}). \qquad (3.8.5)$$

Theorem 3.8.1. *Let (3.8.3) hold. Then*

$$\phi_1(x, \lambda) = \sum_{n=0}^{\infty} \lambda^{-n/2} \phi_{1,n}(x, \lambda), \qquad (3.8.6)$$

$$\phi_2(x, \lambda) = \sum_{n=0}^{\infty} \lambda^{-(n+1)/2} \phi_{2,n}(x, \lambda), \qquad (3.8.7)$$

where
$$\phi_{1,0}(x,\lambda) = \cos x\sqrt{\lambda}, \quad \phi_{2,0}(x,\lambda) = \sin x\sqrt{\lambda},$$
and
$$\phi_{j,n}(x,\lambda) = \int_0^x \sin\left((x-t)\sqrt{\lambda}\right) q(t)\phi_{j,n-1}(t,\lambda)\,dt \quad (n \geq 1,\ j = 1,2).$$

Proof. On iterating (3.8.1) N times, we obtain
$$\phi_1(x,\lambda) = \sum_{n=0}^N \lambda^{-n/2}\phi_{1,n}(x,\lambda) + \lambda^{-(N+1)/2}R_{N+1}(x,\lambda), \tag{3.8.8}$$
where
$$R_n(x,\lambda) = \int_0^x \sin\left((x-t)\sqrt{\lambda}\right) q(t) R_{n-1}(t,\lambda)\,dt \tag{3.8.9}$$
$(1 \leq n \leq N+1)$ and $R_0(x,\lambda) = \phi_1(x,\lambda)$. By (3.8.3), $|\phi_1(x,\lambda)| \leq 2$ and then, by induction on n, we obtain from (3.8.9)
$$|R_n(x,\lambda)| \leq \frac{2}{n!}\left(\int_0^x |q(t)|\,dt\right)^n.$$

Hence $\lambda^{-(N+1)/2}R_{N+1}(x,\lambda) \to 0$ as $N \to \infty$, and (3.8.6) follows from (3.8.8). The proof of (3.8.7) is similar. □

We note that (3.8.6) and (3.8.7) imply in particular that, as $\lambda \to \infty$,
$$\phi_1(x,\lambda) = \cos x\sqrt{\lambda} + \lambda^{-1/2}\int_0^x \sin\left((x-t)\sqrt{\lambda}\right) q(t)\cos t\sqrt{\lambda}\,dt$$
$$+ \lambda^{-1}\int_0^x \sin\left((x-t)\sqrt{\lambda}\right) q(t)\,dt \int_0^t \sin\left((t-u)\sqrt{\lambda}\right) q(u)\cos u\sqrt{\lambda}\,du$$
$$+ O(\lambda^{-3/2}), \tag{3.8.10}$$
and
$$\phi_2(x,\lambda) = \lambda^{-1/2}\sin x\sqrt{\lambda} + \lambda^{-1}\int_0^x \sin\left((x-t)\sqrt{\lambda}\right) q(t)\sin t\sqrt{\lambda}\,dt$$
$$+ \lambda^{-3/2}\int_0^x \sin\left((x-t)\sqrt{\lambda}\right) q(t)\,dt \int_0^t \sin\left((t-u)\sqrt{\lambda}\right) q(u)\sin u\sqrt{\lambda}\,du$$
$$+ O(\lambda^{-2}). \tag{3.8.11}$$

The next theorem gives a first improvement of (3.8.10) and (3.8.11) subject to q being absolutely continuous.

3.8. Asymptotic formulae for solutions

Theorem 3.8.2. Let q be $AC[0, a]$. Then, as $\lambda \to \infty$,

$$\phi_1(x, \lambda) = \cos x\sqrt{\lambda} + \frac{1}{2} \lambda^{-1/2} Q(x) \sin x\sqrt{\lambda}$$
$$+ \frac{1}{4} \lambda^{-1} \left(q(x) - q(0) - \frac{1}{2} Q^2(x)\right) \cos x\sqrt{\lambda} + o(\lambda^{-1}), \qquad (3.8.12)$$

$$\phi_2(x, \lambda) = \lambda^{-1/2} \sin x\sqrt{\lambda} - \frac{1}{2} \lambda^{-1} Q(x) \cos x\sqrt{\lambda}$$
$$+ \frac{1}{4} \lambda^{-3/2} \left(q(x) + q(0) - \frac{1}{2} Q^2(x)\right) \sin x\sqrt{\lambda} + o(\lambda^{-3/2}), \qquad (3.8.13)$$

where
$$Q(x) = \int_0^x q(t)\, dt.$$

Proof. Considering the second term on the right of (3.8.10), we have

$$\int_0^x \sin\left((x-t)\sqrt{\lambda}\right) q(t) \cos t\sqrt{\lambda}\, dt \qquad (3.8.14)$$
$$= \frac{1}{2} \int_0^x \left(\sin x\sqrt{\lambda} + \sin\left((x-2t)\sqrt{\lambda}\right)\right) q(t)\, dt$$
$$= \frac{1}{2} Q(x) \sin x\sqrt{\lambda} + \frac{1}{4} \lambda^{-1/2} \left(q(x) - q(0)\right) \cos x\sqrt{\lambda}$$
$$- \frac{1}{4} \lambda^{-1/2} \int_0^x \cos\left((x-2t)\sqrt{\lambda}\right) q'(t)\, dt. \qquad (3.8.15)$$

The integral on the right here is $o(1)$ as $\lambda \to \infty$ by the Riemann-Lebesgue Lemma. A similar integral to (3.8.14) also occurs in the third term on the right of (3.8.10). Hence, by the above working for (3.8.14), this third term is

$$\frac{1}{2} \lambda^{-1} \int_0^x \sin\left((x-t)\sqrt{\lambda}\right) q(t) Q(t) \sin t\sqrt{\lambda}\, dt + O(\lambda^{-3/2})$$
$$= \frac{1}{4} \lambda^{-1} \int_0^x \left(\cos\left((x-2t)\sqrt{\lambda}\right) - \cos x\sqrt{\lambda}\right) q(t) Q(t)\, dt + O(\lambda^{-3/2})$$
$$= -\frac{1}{8} \lambda^{-1} Q^2(x) \cos x\sqrt{\lambda} + o(\lambda^{-1}),$$

again using the Riemann-Lebesgue Lemma. This together with (3.8.15) proves (3.8.12). The proof of (3.8.13) is similar. □

The asymptotic formulae in this theorem can be improved further if q has derivatives of higher order. As an example, for reference later in the chapter, we have

$$\phi_2(x, \lambda) = \frac{\sin x\sqrt{\lambda}}{\sqrt{\lambda}} \left(1 + \sum_{n=1}^M \frac{S_n(x)}{\lambda^n}\right) + \cos x\sqrt{\lambda} \left(\sum_{n=1}^M \frac{T_n(x)}{\lambda^n}\right) + o(\lambda^{-M-1/2})$$
$$\qquad (3.8.16)$$

if $q^{(2M-2)}$ is absolutely continuous, where S_n and T_n do not depend on λ. Formal substitution of (3.8.16) into (3.2.4) yields recurrence formulae from which S_n and T_n can be determined in terms of q. The details rapidly become cumbersome, but we mention here that

$$4S_1(x) = q(x) + q(0) - \frac{1}{2}Q^2(x) \tag{3.8.17}$$

as in (3.8.13), and

$$\begin{aligned}16S_2(x) =& -(q''(x) + q''(0)) + 2\left(q^2(x) + q^2(0)\right) \\&+ (q(x) + q(0))\left(q(0) - \frac{1}{2}Q^2(x)\right) + \frac{1}{2}\left(q^2(x) - q^2(0)\right) \\&+ Q(x)(q'(x) - q'(0)) - (RQ)(x) + \frac{1}{24}Q^4(x),\end{aligned} \tag{3.8.18}$$

where

$$R(x) = \int_0^x q^2(t)\, dt.$$

3.9 Absence of instability intervals

We have seen in sections 1.7 and 3.7 that, for the Mathieu equation, all the instability intervals are non-empty, that is, $\lambda_{2m+2} > \lambda_{2m+1}$ and $\mu_{2m+1} > \mu_{2m}$ for all $m \geq 0$. This property also holds more widely for periodic equations

$$y'' + (\lambda - q)y = 0 \tag{3.9.1}$$

in a certain general sense which we shall specify in section 3.12. In this section and the next, however, we consider the inverse problem of determining the special nature of q given that only a finite number of the instability intervals are non-empty.

The method here uses the eigenvalue problem comprising (3.9.1), considered to hold in $[\tau, \tau + a]$, and the Dirichlet boundary conditions

$$y(\tau) = y(\tau + a) = 0, \tag{3.9.2}$$

where τ is regarded as a variable parameter. This problem was introduced in Theorem 2.4.5 and the eigenvalues are denoted by $\Lambda_{\tau,n}$ and, simply, Λ_n when $\tau = 0$. We denote by $\phi_{\tau,j}(x, \lambda)$ ($j = 1, 2$) the solutions of (3.9.1) which satisfy the initial conditions

$$\phi_{\tau,1}(\tau, \lambda) = 1, \quad \phi'_{\tau,1}(\tau, \lambda) = 0; \quad \phi_{\tau,2}(\tau, \lambda) = 0, \quad \phi'_{\tau,2}(\tau, \lambda) = 1. \tag{3.9.3}$$

By matching initial values at $x = \tau$, we obtain

$$\left.\begin{aligned}\phi_{\tau,1}(x, \lambda) &= \phi'_2(\tau, \lambda)\phi_1(x, \lambda) - \phi'_1(\tau, \lambda)\phi_2(x, \lambda), \\\phi_{\tau,2}(x, \lambda) &= -\phi_2(\tau, \lambda)\phi_1(x, \lambda) + \phi_1(\tau, \lambda)\phi_2(x, \lambda).\end{aligned}\right\} \tag{3.9.4}$$

3.9. Absence of instability intervals

When $\tau = 0$ we have of course $\phi_{\tau,j}(x, \lambda) = \phi_j(x, \lambda)$. One relation which follows immediately from (3.9.4) and which we require later is

$$\frac{d}{d\tau}\phi_{\tau,2}(\tau + a, \lambda) = -\phi_{\tau,1}(\tau + a, \lambda) + \phi'_{\tau,2}(\tau + a, \lambda). \tag{3.9.5}$$

The following theorem is the first step on the way to determining the special nature of q.

Theorem 3.9.1. *If all but a finite number of the instability intervals of* (3.9.1) *are absent, then $q(x)$ coincides with an infinitely differentiable function a.e.*

Proof. We recall Theorem 2.4.5 which shows that, as τ varies, each $\Lambda_{\tau,n}$ varies throughout an instability interval of (3.9.1). Hence, if an instability interval is absent, the corresponding $\Lambda_{\tau,n}$ is independent of τ. By hypotheses, therefore, only a finite number of the $\Lambda_{\tau,n}$ depend on τ. We denote these by $\Lambda_\tau^{(m)}$ and, when $\tau = 0$, by $\Lambda^{(m)}$ ($m = 1, 2, \ldots, N$).

By (3.9.2) and the initial conditions for $\phi_{\tau,2}(x.\lambda)$ in (3.9.3), the $\Lambda_{\tau,n}$ are the values of λ such that $\phi_{\tau,2}(\tau + a, \lambda) = 0$, that is, they are the zeros of $\phi_{\tau,2}(\tau + a, \lambda)$ considered as a function of λ, these zeros being simple. Now consider, as functions of λ, $f(\lambda)$ and $g(\lambda)$ defined by

$$f(\lambda) = \phi_{\tau,2}(\tau + a, \lambda) \prod_{m=1}^{N} (\lambda - \Lambda^{(m)}),$$

$$g(\lambda) = \phi_2(a, \lambda) \prod_{m=1}^{N} (\lambda - \Lambda_\tau^{(m)}),$$

where λ is regarded as a complex variable. Then $f(\lambda)$ and $g(\lambda)$ are integral functions of λ of order $\frac{1}{2}$ and they have the same zeros, each zero, though not necessarily simple, being of the same order for $f(\lambda)$ and $g(\lambda)$. Hence, by the Hadamard Factorisation Theorem, there is a number $M(\tau)$, independent of λ but possibly depending on τ, such that $f(\lambda) = M(\tau)g(\lambda)$. Now $\phi_2(a, \lambda)$ has the asymptotic form

$$\lambda^{-1/2} \sin a\sqrt{\lambda} + O(\lambda^{-1})$$

by (3.8.5) when λ is real and $\lambda \to \infty$, and it is easily verified (cf. (3.9.18) below) that $\phi_{\tau,2}(\tau + a, \lambda)$ has the same asymptotic form. From this it follows that $M(\tau) = 1$, and so $f(\lambda) = g(\lambda)$. Thus

$$\phi_{\tau,2}(\tau + a, \lambda) = \phi_2(a, \lambda)h(\tau, \lambda), \tag{3.9.6}$$

where

$$h(\tau, \lambda) = \prod_{1}^{N} (\lambda - \Lambda_\tau^{(m)})/(\lambda - \Lambda^{(m)}). \tag{3.9.7}$$

We note that, by (3.9.4), $\phi_{\tau,2}(\tau + a, \lambda)$ is twice differentiable with respect to τ and hence, by (3.9.6), the same is true of $h(\tau, \lambda)$. On differentiating (3.9.6) with respect to τ and using (3.9.5), we obtain

$$\phi'_{\tau,2}(\tau + a, \lambda) - \phi_{\tau,1}(\tau + a, \lambda) = \phi_2(a, \lambda)\frac{\partial h}{\partial \tau}(\tau, \lambda). \tag{3.9.8}$$

Hence, with $\tau = 0$,

$$\phi'_2(a, \lambda) - \phi_1(a, \lambda) = \phi_2(a, \lambda)\frac{\partial h}{\partial \tau}(0, \lambda). \tag{3.9.9}$$

When λ is real and $\lambda \to \infty$, it follows from this and (3.8.5) and (3.9.7) that

$$\phi'_2(a, \lambda) - \phi_1(a, \lambda) = O(\lambda^{-3/2}). \tag{3.9.10}$$

Again on differentiating (3.9.6) twice with respect to τ, and proceeding similarly, we obtain

$$(\lambda - q(0)) \phi_2(a, \lambda) + \phi'_1(a, \lambda) = O(\lambda^{-3/2}). \tag{3.9.11}$$

Our purpose now is to show that (3.9.10) and (3.9.11) imply that $q(x)$ coincides with an absolutely continuous function a.e. We let $\lambda \to \infty$ through real values and we substitute from (3.8.10) and (3.8.11) into the left-hand side of (3.9.10). This gives

$$\int_0^a \sin\left((2t - a)\sqrt{\lambda}\right) q(t)\, dt$$
$$+ \lambda^{-1/2} \int_0^a q(t)\,dt \int_0^t \sin\left((t + u - a)\sqrt{\lambda}\right) \sin\left((t - u)\sqrt{\lambda}\right) q(u)\, du$$
$$= O(\lambda^{-1}). \tag{3.9.12}$$

It is convenient at this point to impose the condition that $q(x)$ has mean-value zero, i.e.,

$$\int_0^a q(x)\,dx = 0.$$

This condition can be satisfied without altering (3.9.1) by merely adding a suitable constant to both λ and $q(x)$. The consequence is that the integral

$$Q(x) := \int_0^x q(t)\,dt$$

also has period a.

On integrating by parts in the u-integral, the second term on the left of (3.9.12) can be written

$$\int_0^a q(t)\,dt \int_0^t Q(u) \sin\left((2u - a)\sqrt{\lambda}\right) du = -\int_0^a Q^2(u) \sin\left((2u - a)\sqrt{\lambda}\right) du$$

3.9. Absence of instability intervals

after changing the order of integration. Thus (3.9.12) gives

$$\int_0^a (q(t) - Q^2(t)) \sin\left((2t-a)\sqrt{\lambda}\right) dt = O(\lambda^{-1}). \tag{3.9.13}$$

Similar reasoning applies to (3.9.11), and the corresponding result is

$$\int_0^a (q(t) - Q^2(t)) \cos\left((2t-a)\sqrt{\lambda}\right) dt - q(0)\lambda^{-1/2} \sin a\sqrt{\lambda} = O(\lambda^{-1}). \tag{3.9.14}$$

On taking $\lambda = n^2\pi^2/a^2$, where n is an integer, in (3.9.13) and (3.9.14), we obtain

$$A_n = O(n^{-2}), \quad B_n = O(n^{-2}) \tag{3.9.15}$$

as $n \to \infty$, A_n and B_n being the real Fourier coefficients of $q(x) - Q^2(x)$ referred to the interval $[0, a]$. By (3.9.15), $q(x) - Q(x)$ has a uniformly convergent Fourier series, and it is therefore equal to the sum of its Fourier series a.e. Further, since $\sum n^2(A_n^2 + B_n^2)$ is convergent, the Riesz-Fisher Theorem shows that the differentiated Fourier series is the Fourier series of an $L^2(0, a)$ function, $q_0(x)$ say. Now the Fourier series of $q_0(x)$ can be integrated term by term to give the sum

$$\int_0^x q_0(t) dt,$$

and hence we have

$$q(x) - Q^2(x) + C = \int_0^x q_0(t) dt \tag{3.9.16}$$

almost everywhere, where C is a constant. By altering the values of $q(x)$ on a set of measure zero, we take it that (3.9.16) holds for all x. Since $Q)x)$ is absolutely continuous, it follows from (3.9.16) that $q(x)$ is also absolutely continuous.

The final stage of the proof is to conclude that $q(x)$ is in fact infinitely differentiable. With $q(x)$ now known to be absolutely continuous, we can use Theorem 3.8.2. Since $q(a) = q(0)$ and $Q(a) = 0$, (3.8.13) gives

$$\phi_2(a, \lambda) = \lambda^{-1/2} \sin a\sqrt{\lambda} + \tfrac{1}{2}\lambda^{-3/2} q(0) \sin a\sqrt{\lambda} + o(\lambda^{-3/2}). \tag{3.9.17}$$

A similar asymptotic formula holds for $\phi_{\tau,2}(\tau + a, \lambda)$ and this is easily seen to be

$$\phi_{\tau,2}(\tau + a, \lambda) = \lambda^{-1/2} \sin a\sqrt{\lambda} + \tfrac{1}{2}\lambda^{-3/2} q(\tau) \sin a\sqrt{\lambda} + o(\lambda^{-3/2}). \tag{3.9.18}$$

Substituting (3.9.17) and (3.9.18) into (3.9.6) and expanding $h(\tau, \lambda)$ as a series in λ^{-1}, we obtain

$$\sin a\sqrt{\lambda} + \frac{1}{2}\lambda^{-1} \sin a\sqrt{\lambda} + o(\lambda^{-1}) = \left(\sin a\sqrt{\lambda} + \frac{1}{2}\lambda^{-1} \sin a\sqrt{\lambda} + o(\lambda^{-1})\right)$$

$$\times \left(1 - \lambda^{-1} \sum_{m=1}^N (\Lambda_\tau^{(m)} - \Lambda^{(m)}) + O(\lambda^{-2})\right).$$

By comparing the terms in λ^{-1}, we obtain

$$q(\tau) - q(0) = -2 \sum_{m=1}^{N} (\Lambda_\tau^{(m)} - \Lambda^{(m)}). \tag{3.9.19}$$

Now, by (3.9.4), $\phi_{\tau,2}(\tau + a, \lambda)$ is twice differentiable with respect to τ. By (3.9.6), the same is true of $h(\tau, \lambda)$ and therefore of each $\Lambda_\tau^{(m)}$. Hence, by (3.9.19), $q''(\tau)$ exists. Then, returning to (3.9.1), all solutions of (3.9.1) are four times differentiable with respect to x. This implies, by (3.9.4), that $\phi_{\tau,2}(\tau + a, \lambda)$ is four times differentiable with respect to τ, and then, as above, (3.9.19) shows that $q^{(4)}(\tau)$ exists. This circular argument can be repeated indefinitely and it shows that $q(\tau)$ is infinitely differentiable, thereby proving the theorem. $\square$

A particular case of the above proof gives the following result.

Theorem 3.9.2. *If all the instability intervals of (3.9.1) are absent, except for the zero-th, then $q(x)$ is a constant a.e.*

Proof. The method of the previous proof applies but with $h(\tau, \lambda)$ replaced by unity in (3.9.6). Then (3.9.19) is replaced by, simply, $q(\tau) - q(0) = 0$, and the theorem follows. $\square$

3.10 Absence of all but N finite instability intervals

The analysis in the previous section can be taken further to show, not only that q is infinitely differentiable, but that q has a property related more specifically to the number N in (3.9.7). This property is that q satisfies a differential equation of order $2N$.

We require some preliminary results concerning the function $\phi_{\tau,2}(x, \lambda)$ in (3.9.3) and (3.9.4). The first is that

$$\psi_\tau(x, \lambda) := \phi_{\tau,2}(x + \tau, \lambda) \tag{3.10.1}$$

satisfies the differential equation

$$y''(x) + (\lambda - q(x + \tau))\, y(x) = 0 \tag{3.10.2}$$

and the initial conditions

$$\psi_\tau(0, \lambda) = 0, \quad \psi'_\tau(0, \lambda) = 1,$$

from which it follows that ψ_τ is infinitely differentiable as a function of τ. A consequence of (3.10.2) is that the asymptotic formula (3.9.18) can be developed

3.10. Absence of all but N finite instability intervals

further, along the lines of (3.8.16). Since we now have $q(x+\tau)$ in place of $q(x)$, the result corresponding to (3.8.16) is

$$\psi_\tau(x,\lambda) = \frac{\sin x\sqrt{\lambda}}{\sqrt{\lambda}}\left(1+\sum_{n=1}^{M}\frac{S_n(x,\tau)}{\lambda^n}\right) + \cos x\sqrt{\lambda}\left(\sum_{n=1}^{M}\frac{T_n(x,\tau)}{\lambda^n}\right) + O(\lambda^{-M-1}) \tag{3.10.3}$$

as $\lambda \to \infty$, where S_n and T_n do not depend on λ. This formula holds for any integer $M \geq 0$ since q is now infinitely differentiable.

Since we shall be using (3.9.6) again, we are interested in the value $x=a$ in (3.10.3) and we also take $M = N+1$. The focus of our attention will be the functions $S_n(a,\tau)$, which we abbreviate to $S_n(\tau)$. Thus, to eliminate $T_n(a,\tau)$ from the discussion we also choose

$$\sqrt{\lambda} = (2k+\tfrac{1}{2})\pi/a \tag{3.10.4}$$

where k is an integer. With such values of λ, (3.10.3) becomes

$$\psi_\tau(a,\lambda) = \frac{1}{\sqrt{\lambda}}\sum_{n=0}^{N+1}\frac{S_n(\tau)}{\lambda^n} + O(\lambda^{-N-2}), \tag{3.10.5}$$

where $S_0(\tau) = 1$. The nature of the S_n is determined in the following two lemmas.

Lemma 3.10.1. *The S_n satisfy the recurrence relation*

$$S'_{n+1} = -\frac{1}{4}S'''_n + qS'_n + \frac{1}{2}q'S_n. \tag{3.10.6}$$

Proof. By (3.10.1) and (3.9.4), we have

$$\psi_\tau(a,\lambda) := \phi_{\tau,2}(a+\tau,\lambda)$$
$$= -\phi_2(\tau,\lambda)\phi_1(a+\tau,\lambda) + \phi_1(\tau,\lambda)\phi_2(a+\tau,\lambda).$$

On differentiating three times with respect to τ and using the fact that ϕ_1 and ϕ_2 are solutions of (3.9.1), we obtain

$$d^3\psi_\tau(a,\lambda)/d\tau^3 = 2q'(\tau)\psi_\tau(a,\lambda) + 4(q(\tau)-\lambda)\,d\psi_\tau(a,\lambda)/d\tau.$$

We substitute (3.10.5) into this equation, and (3.10.6) follows when we consider the terms involving λ^{-n} on each side. There is of course the question of differentiating the O-term in (3.10.5) but we omit the straightforward but lengthy proof that it remains as $O(\lambda^{-N-2})$. □

Lemma 3.10.2. *S_n has the form*

$$S_n = (-1)^{n-1}\frac{1}{2^{2n-1}}q^{(2n-2)} + f_n(q,\ldots,q^{(2n-4)}) \tag{3.10.7}$$

where f_n is a polynomial of degree at most n.

Proof. The main part of the proof is to deduce from (3.10.6) that S_n is a polynomial in q and its derivatives. The more precise form (3.10.7) is then an immediate consequence of (3.10.6). We now prove by induction the following statement.

($\star$)   Both S_n and $\int_0^x qS'_n$ are polynomials in q and its derivatives.

Since $S_0 = 1$ and $S_1 = \frac{1}{2}q + (const.)$ by (3.10.6), ($\star$) is true for $n = 1$. Now assume that ($\star$) is true for $n = 1, \ldots, J$. Then

$$S_{J+1} = -\frac{1}{4}S''_J + \frac{1}{2}qS_J + \frac{1}{2}\int_0^x qS'_J + (const),$$

and S_{J+1} is therefore a polynomial in q and its derivatives. To deal with $\int_0^x qS'_{J+1}$, consider for $n \leq J$,

$$\int_0^x S_n S'_{J+1} = -\frac{1}{4}\int_0^x S_n S'''_J + \int_0^x qS_n S'_J + \frac{1}{2}\int_0^x q'S_n S_J$$

$$= -\frac{1}{4}\int_0^x S''_n S'_J + \frac{1}{4}[S'_n S'_J - S_n S''_J]\Big|_0^x + \int_0^x qS_n S'_J$$

$$- \frac{1}{2}\int_0^x (\int_0^t q'S_n)S'_J - \frac{1}{2}S_J \int_0^x qS'_n + \frac{1}{2}[qS_n]\Big|_0^x S_J$$

after integrations by parts. Hence, denoting by $\{\ldots\}$ any expression which is a polynomial in q and its derivatives, we have

$$\int_0^x S_n S'_{J+1} = \int_0^x \left(-\frac{1}{4}S''_n + qS_n - \frac{1}{2}\int_0^t q'S_n\right)S'_J + \{\ldots\} = \int_0^x S_{n+1} S'_J + \{\ldots\}$$

by (3.10.6). Thus, for $n + k \leq J$,

$$\int_0^x S_n S'_{J+1} = \int_0^x S_{n+k} S'_{J-k+1} + \{\ldots\}.$$

In particular, with $n = 1$ we have

$$\int_0^x qS'_{J+1} = \int_0^x S_{k+1} S'_{J-k+1} + \{\ldots\} \quad (k+1 \leq J). \tag{3.10.8}$$

When J is even ($= 2k$, say), (3.10.8) gives

$$\int_0^x qS'_{J+1} = \int_0^x S_{k+1} S'_{k+1} + \{\ldots\}$$

$$= \tfrac{1}{2}S^2_{k+1} + \{\ldots\} = \{\ldots\},$$

3.10. Absence of all but N finite instability intervals

as required. When J is odd $(= 2k+1$, say), (3.10.8) gives

$$\int_0^x qS'_{J+1} = \int_0^x S_{k+1} S'_{k+2} + \{\dots\}$$
$$= \int_0^x S_{k+1}\left(-\frac{1}{4}S'''_{k+1} + qS'_{k+1} + \frac{1}{2}q'S_{k+1}\right) + \{\dots\}$$
$$= \frac{1}{8}S'^2_{k+1} + \frac{1}{2}qS^2_{k+1} + \{\dots\} = \{\dots\}$$

again as required. This completes the induction proof of $(\star)$, and then (3.10.7) follows as mentioned above. $\square$

When $n \geq 2$, (3.10.6) implies a little more than (3.10.7), viz.

$$S_n = (-1)^{n-1}\frac{1}{2^{2n-1}}q^{(2n-2)} + \frac{(2n)!}{2^{2n}(n!)^2}q^n + g_n(q, \dots q^{(2n-4)}), \qquad (3.10.9)$$

where g_n is a polynomial of degree at most $n-1$.

We can now state and prove the main result of this section.

Theorem 3.10.3. *Let all but N of the finite instability intervals of (3.9.1) be absent. Then $q(x)$ satisfies a differential equation of the form*

$$q^{(2N)} = (-1)^N \frac{(2N+2)!}{2((N+1)!)^2} q^{N+1} + g(q, q', \dots, q^{(2N-2)}), \qquad (3.10.10)$$

where g is a polynomial of degree at most N.

Proof. We return to (3.9.6) and (3.9.7) with the choice $\sqrt{\lambda} = (2k + \frac{1}{2})\pi/a$ made in (3.10.4). By (3.10.5), we can write (3.9.6) as

$$\Big(\sum_{n=0}^{N+1} \frac{S_n(\tau)}{\lambda^n}\Big)\Big(1 + \frac{a_1(0)}{\lambda} + \dots + \frac{a_N(0)}{\lambda^N}\Big)$$
$$= \Big(\sum_{n=0}^{N+1} \frac{S_n(0)}{\lambda^n}\Big)\Big(1 + \frac{a_1(\tau)}{\lambda} + \dots + \frac{a_N(\tau)}{\lambda^N}\Big) + O(\lambda^{-N-3/2}).$$

Equating terms in λ^{-n} $(n = 1, \dots, N)$ on each side, we obtain

$$S_n(\tau) + a_1(0)S_{n-1}(\tau) + \dots + a_{n-1}(0)S_1(\tau) + a_n(0)$$
$$= S_n(0) + a_1(\tau)S_{n-1}(0) + \dots + a_{n-1}(\tau)S_1(0) + a_n(\tau),$$

from which it follows that each $a_n(\tau)$ in turn is a linear combination of $S_0(\tau), S_1(\tau), \dots, S_n(\tau)$. Then, equating terms in λ^{-N-1}, we also have

$$S_{N+1}(\tau) + a_1(0)S_N(\tau) + \dots + a_N(0)S_1(\tau)$$
$$= S_{N+1}(0) + a_1(\tau)S_N(0) + \dots + a_N(\tau)S_1(0),$$

from which it follows that $S_{N+1}(\tau)$ is a linear combination of $S_N(\tau), \ldots S_0(\tau)$:

$$S_{N+1}(\tau) = \sum_{n=0}^{N} c_n S_n(\tau) \tag{3.10.11}$$

for some constants c_n. Then (3.10.10) follows immediately from (3.10.9). □

In addition to (3.10.9), we note here the first few solutions S_n of (3.10.6), starting with $S_0 = 1$. Thus

$$S_1 = \tfrac{1}{2} q, \qquad S_2 = -\tfrac{1}{8}(q'' - 3q^2), \tag{3.10.12}$$
$$S_3 = \tfrac{1}{32}(q^{(4)} - 5q'^2 - 10qq' + 10q^3).$$

There would normally be constants of integration on the right of these equations, but these constants are all zero and we refer to Goldberg [68],[69] for the details to support this statement. The formulae for S_1 and S_2 also follow from (3.8.17) and (3.8.18) when we put $x = a$ and replace $q(x)$ by $q(x + \tau)$. In the notes at the end of this chapter, we indicate briefly that the S_n are related to a hierarchy of KdV equations.

Finally in this section we note that, when $N = 1$, (3.10.11) and (3.10.12) give

$$q'' = 3q^2 + k_1 q + k_2$$

with constants k_1 and k_2, making q a multiple (namely 2) of a Weierstrass elliptic function together with an additive constant.

3.11 Absence of odd instability intervals

Another situation in which a special property of q is implied by the absence of instability intervals occurs when it is the odd intervals that vanish. Thus we suppose in this section that

$$\mu_{2m} = \mu_{2m+1} \quad (m \geq 0), \tag{3.11.1}$$

that is, the zeros of $D(\lambda) + 2$ are all double and the solutions of (3.9.1) corresponding to (3.11.1) all have semi-period a.

Theorem 3.11.1. *Let (3.11.1) hold. Then q has period $\tfrac{1}{2}a$.*

Proof. We again use the solution $\phi_{\tau,2}(x, \lambda)$ of (3.9.1) which satisfies the initial conditions in (3.9.3), and the proof depends on comparing the two functions

$$f(\lambda) := \phi_{\tau,2}(\tau - \tfrac{1}{2}a, \lambda) + \phi_{\tau,2}(\tau + \tfrac{1}{2}a, \lambda) \tag{3.11.2}$$

and

$$g(\lambda) := \sqrt{D(\lambda) + 2}$$

3.11. Absence of odd instability intervals

with the parameter λ taken to be complex. Then $f(\lambda)$ is an integral function of λ and, because of (3.11.1), $g(\lambda)$ is also an integral function whose zeros are all simple. Further, $f(\lambda)/g(\lambda)$ is an integral function because, at a zero μ_{2m} of $g(\lambda)$, $\phi_{\tau,2}(x,\mu_{2m})$ has semi-period a, making μ_{2m} also a zero of $f(\lambda)$.

We require the asymptotics of $f(\lambda)$ and $g(\lambda)$ when $|\lambda|$ is large, and we write $\sqrt{\lambda} = \mu + i\nu$. To deal first with $D(\lambda)$ ($= \phi_1(a,\lambda) + \phi_2'(a,\lambda)$) we use (3.8.1) and (3.8.2). Since $|\cos z| \le \exp|\operatorname{im} z|$ (and similarly for $\sin z$), it follows from (3.8.1) that $|\phi_1(x,\lambda)| \le 2\exp|x\nu|$ if $|\lambda|^{\frac{1}{2}} \ge 2\int_0^a |q(t)|\,dt$, by similar reasoning to that leading to (3.8.3). Then, again by (3.8.1),

$$\phi_1(x,\lambda) = \cos x\sqrt{\lambda} + O\big(|\lambda|^{-\frac{1}{2}} \exp|x\nu|\big).$$

The same asymptotic expression for $\phi_2'(x,\lambda)$ follows from (3.8.2), and hence

$$D(\lambda) + 2 = 4\cos^2\left(\tfrac{1}{2}a\sqrt{\lambda}\right) + O\big(|\lambda|^{-\frac{1}{2}}\exp(a|\nu|)\big). \tag{3.11.3}$$

If, in particular,

$$\operatorname{Re}\sqrt{\lambda} = 4N\pi/a \tag{3.11.4}$$

where N is an integer, $\cos\left(\tfrac{1}{2}a\sqrt{\lambda}\right) = \cosh\left(\tfrac{1}{2}a\nu\right) \ge \tfrac{1}{2}\exp\left(\tfrac{1}{2}a\nu\right)$, and then (3.11.3) gives

$$g(\lambda) = 2\cos\left(\tfrac{1}{2}a\sqrt{\lambda}\right)\left(1 + O(|\lambda|^{-\frac{1}{2}})\right). \tag{3.11.5}$$

Turning now to $\phi_{\tau,2}(x,\lambda)$, we have corresponding to (3.8.2)

$$\phi_{\tau,2}(x,\lambda) = \lambda^{-\frac{1}{2}} \sin\left((x-\tau)\sqrt{\lambda}\right) + \lambda^{-\frac{1}{2}} \int_\tau^x \sin\left((x-t)\sqrt{\lambda}\right) q(t)\phi_\tau(t,\lambda)\,dt. \tag{3.11.6}$$

It again follows that $|\phi_{\tau,2}(x,\lambda)| \le 2|\lambda|^{-\frac{1}{2}}\exp|(x-\tau)\nu|$, and iteration of (3.11.6) gives

$$\phi_{\tau,2}(x,\lambda) = \lambda^{-\frac{1}{2}} \sin\left((x-\tau)\sqrt{\lambda}\right) - \tfrac{1}{2}\lambda^{-1}\cos\left((x-\tau)\sqrt{\lambda}\right)\int_\tau^x q(t)\,dt$$
$$+ \tfrac{1}{2}\lambda^{-1} \int_\tau^x \cos\left((x+\tau-2t)\sqrt{\lambda}\right) q(t)\,dt + O\big(|\lambda|^{-\frac{3}{2}}\exp|(x-\tau)\nu|\big).$$

The second integral on the right is $o\big(\exp|(x-\tau)\nu|\big)$ by the same argument used to prove the Riemann-Lebesgue Lemma. Hence, by (3.11.2),

$$f(\lambda) = -\tfrac{1}{2}\lambda^{-1}\cos\left(\tfrac{1}{2}a\sqrt{\lambda}\right)\left(\int_\tau^{\tau-\frac{1}{2}a} + \int_\tau^{\tau+\frac{1}{2}a}\right) q(t)\,dt + o\big(|\lambda|^{-1}\exp\left(\tfrac{1}{2}a|\nu|\right)\big). \tag{3.11.7}$$

In particular, when (3.11.4) holds,

$$f(\lambda) = -\frac{1}{2}\lambda^{-1}\cos\left(\frac{1}{2}a\sqrt{\lambda}\right)\left(\left(\int_\tau^{\tau-\frac{1}{2}a} + \int_\tau^{\tau+\frac{1}{2}a}\right)q(t)\,dt + o(1)\right). \quad (3.11.8)$$

We now introduce a large closed contour Γ_N in the λ-plane as follows. The upper half of Γ_N corresponds to the quarter square in the $\sqrt{\lambda}$-plane formed by the lines

$$\operatorname{Re}\sqrt{\lambda} = 4N\pi/a \quad (0 \le \operatorname{Im}\sqrt{\lambda} \le 4N\pi/a), \quad (3.11.9)$$

$$\operatorname{Im}\sqrt{\lambda} = 4N\pi/a \quad (0 \le \operatorname{Re}\sqrt{\lambda} \le 4N\pi/a), \quad (3.11.10)$$

and then Γ_N is symmetrical about the real axis. Here N is a large positive integer. When (3.11.9) holds we have

$$f(\lambda)/g(\lambda) = O(|\lambda|^{-1}) \quad (3.11.11)$$

by (3.11.4), (3.11.5) and (3.11.8). When (3.11.10) holds we have

$$\tfrac{1}{4}\exp 2N\pi/a \le \left|\cos\tfrac{1}{2}a\sqrt{\lambda}\right| \le \exp 2N\pi/a$$

for large N, and hence (3.11.11) holds again, by (3.11.3) and (3.11.7).

Thus $f(\lambda)/g(\lambda)$ is bounded on the sequence Γ_N of contours receding to infinity. Then, by Liouville's Theorem, $f(\lambda)/g(\lambda)$ is a constant, the constant being zero by (3.11.11). Hence $f(\lambda) \equiv 0$ and, by (3.11.7),

$$\left(\int_\tau^{\tau-\frac{1}{2}a} + \int_\tau^{\tau+\frac{1}{2}a}\right)q(t)\,dt = 0.$$

Differentiation with respect to τ gives

$$q(\tau - \tfrac{1}{2}a) + q(\tau + \tfrac{1}{2}a) - 2q(\tau) = 0.$$

Since q has period a, we obtain $q(\tau + \tfrac{1}{2}a) = q(\tau)$, proving the theorem. □

Corollary 3.11.2. *Let K (≥ 1) be an integer and let all instability intervals I_n vanish when $2^K \nmid n$. Then q has period $a/2^K$.*

Proof. The theorem is the case $K = 1$. Now consider $K = 2$, so that the vanishing I_n are $I_1, I_2, I_3, I_5, \ldots$. Then certainly q has period $a/2$ and we consider the semi-periodic eigenvalues $\tilde{\mu}_n$ for the interval $(0, \tfrac{1}{2}a)$. By Theorem 2.5.1 the corresponding eigenfunctions $\tilde{\xi}_{2m}$ and $\tilde{\xi}_{2m+1}$ have $2m+1$ zeros in $[0, \tfrac{1}{2}a)$, and hence $4m+2$ zeros in $[0, a)$. Hence $\tilde{\mu}_{2m} = \lambda_{4m+1}$ and $\tilde{\mu}_{2m+1} = \lambda_{4m+2}$. But λ_{4m+1} and λ_{4m+2} are the end-points of I_{4m+2}, which vanishes. Thus $\tilde{\mu}_{2m} = \tilde{\mu}_{2m+1}$, and hence q has period $a/4$ by Theorem 3.11.1 applied to $(0, \tfrac{1}{2}a)$. Repetition of the argument establishes the corollary for general K. □

3.12 All instability intervals non-vanishing

In the results of the preceding sections, the absence of infinitely many instability intervals implies a corresponding special property of q. The opposite question then is whether all instability intervals are present for most (in some sense) potentials q. The setting for dealing with this question is a normed space which, for our purposes, we take to be the space $\mathcal{F}$ of essentially bounded a-periodic functions f with
$$\|f\|_\infty = \operatorname{ess\,sup}_{x \in \mathbb{R}} |f(x)|.$$
We recall some basic topological definitions. A subset S of $\mathcal{F}$ is said to be *nowhere dense* if the closure $\overline{S}$ has empty interior. A subset T of $\mathcal{F}$ is said to be *meagre* if T is a countable union of nowhere dense sets. A meagre set is also called a *set of the first (Baire) category* in $\mathcal{F}$. A subset $\mathcal{U}$ of $\mathcal{F}$ is said to contain *Baire-almost all* elements of $\mathcal{F}$ if the complement $\mathcal{F} \setminus \mathcal{U}$ is meagre.

Theorem 3.12.1. *All instability intervals for the Hill equation*
$$y'' + (\lambda - q) y = 0$$
and for the periodic Dirac system
$$-i\sigma_2 u' + \sigma_3 u + q u = \lambda u$$
are non-vanishing for Baire-almost all $q \in \mathcal{F}$.

We remark that we have made the choices $p_1 = 1$, $p_2 = 0$ in the Dirac system; this corresponds to the case of a particle of constant mass. The statement of Theorem 3.12.1 is false if $p_1 = 0$ as well.

Proof. We first consider a fixed integer $n \in \mathbb{N}$ for the Hill case, or $n \in \mathbb{Z}$ for the Dirac case, and denote by Q_n the set of all $q \in \mathcal{F}$ for which
$$\lambda_{2m+1} < \lambda_{2m+2} \qquad (n = 2m + 2) \tag{3.12.1}$$
or
$$\mu_{2m} < \mu_{2m+1} \qquad (n = 2m + 1).$$
In the first part of the proof, we show that Q_n is open in $\mathcal{F}$ and, in the second part, that Q_n is dense in $\mathcal{F}$. It then follows that the complement $\mathcal{F} \setminus Q_n$ has empty interior and is therefore nowhere dense. Thus the set $\cup_n (\mathcal{F} \setminus Q_n)$ of all potentials for which at least one instability interval vanishes is meagre, and this is what the theorem states.

Part 1. The proof that Q_n is open is the same for even and odd n, and we take (3.12.1). Let $q \in Q_n$ with
$$\lambda_{2m+2} - \lambda_{2m+1} =: 2\delta > 0$$

and let q_1 be any function in $\mathcal{F}$ with $||q_1||_\infty < \delta$.

Then, applying Theorem 4.2.5 with $\mu = (\lambda_{2m+1} + \lambda_{2m-1})/2$, we see that the periodic eigenvalues λ'_{2m+1} and λ'_{2m+2} corresponding to the potential $q+q_1$ satisfy

$$\lambda'_{2m+2} - \lambda'_{2m+1} > 2(\delta - ||q_1||_\infty) > 0.$$

Thus the δ-neighbourhood of q lies in Q_n, showing that Q_n is open.

Part 2. To prove that Q_n is dense in $\mathcal{F}$, we have to show that any $q \in \mathcal{F}\backslash Q_n$ is the limit of elements of Q_n. Suppose then that, contrary to (3.12.1) for example,

$$\lambda_{2m+1} = \lambda_{2m+2} =: \Lambda \tag{3.12.2}$$

say, and let $\Phi = \begin{pmatrix} u_1 & v_1 \\ u_2 & v_2 \end{pmatrix}$ be the canonical fundamental system of the differential equation

$$u' = J(B + \Lambda W)u.$$

We note that $W = \begin{pmatrix} 1 & 0 \\ 0 & 0 \end{pmatrix}$ for the Hill equation and $W = I$ for the Dirac system. Denote by $D(\Lambda, \epsilon)$ the discriminant of the perturbed a-periodic differential equation

$$u' = J(B + \Lambda W + \epsilon \tilde{W})u, \tag{3.12.3}$$

where $\epsilon \in \mathbb{R}$ is a small parameter and $\tilde{W} = rW$, with $r \in \mathcal{F}$ to be chosen later. In particular,

$$D(\Lambda, 0) = 2. \tag{3.12.4}$$

Now following the proof of Theorem 1.6.1 with $\epsilon \tilde{W}$ taking the role of λW, and using (1.6.5) with $\Phi(a) = I$ due to the coexistence, we obtain

$$\frac{\partial D}{\partial \epsilon}(\Lambda, 0) = 0. \tag{3.12.5}$$

Also, from (1.6.10),

$$\frac{1}{2}\frac{\partial^2 D}{\partial \epsilon^2}(\Lambda, 0) = \left(\int_0^a u^T \tilde{W} v\right)^2 - \left(\int_0^a v^T \tilde{W} v\right)\left(\int_0^a u^T \tilde{W} u\right). \tag{3.12.6}$$

Now

$$D(\Lambda, \epsilon) = D(\Lambda, 0) + \epsilon \frac{\partial D}{\partial \epsilon}(\Lambda, 0) + \frac{1}{2}\epsilon^2 \frac{\partial^2 D}{\partial \epsilon^2}(\Lambda, 0) + O(\epsilon^3)$$
$$= 2 + \frac{1}{2}\epsilon^2 \frac{\partial^2 D}{\partial \epsilon^2}(\Lambda, 0) + O(\epsilon^3)$$

by (3.12.4) and (3.12.5). If therefore

$$\frac{\partial^2 D}{\partial \epsilon^2}(\Lambda, 0) > 0, \tag{3.12.7}$$

3.12. All instability intervals non-vanishing

we have $D(\Lambda, \epsilon) > 2$ for small non-zero ϵ, and hence there is a non-vanishing instability interval containing Λ. Then $q - \epsilon r \in Q_n$ for small ϵ, as required for the proof of Part 2.

Thus it remains to choose the function $r : [0, a] \to \mathbb{R}$ to make (3.12.7) hold. We note that the Cauchy-Schwarz inequality gives

$$I := \int_0^a (u^T W u)(v^T W v) < \left(\int_0^a (u^T W u)^2 \int_0^a (v^T W v)^2 \right)^{\frac{1}{2}} \quad (3.12.8)$$

with strict inequality provided we show that $u^T W u$ and $v^T W v$ are linearly independent.

For Hill's equation the linear independence is clear, since $(u^T W u)(0) = 1$ and $(v^T W v)(0) = 0$, but $v^T W v \not\equiv 0$.

For the Dirac equation, we note that $u^T W u = u^T u = \rho_u^2$ and similarly for v and, since $\rho_u^2(0) = \rho_v^2(0)$, u and v are either linearly independent or identical. In the latter case, (2.2.8) with $p_1 = 1$ and $p_2 = 0$ gives

$$\sin 2\theta_u = \sin 2\theta_v,$$

both vanishing at 0, and hence both $\cos 2\theta_u = \cos 2\theta_v$ and $\theta_u' = \theta_v'$ in any interval containing 0 in which $\cos 2\theta_u$ (and hence $\cos 2\theta_v$) does not vanish. But then (2.2.7) shows that $\Lambda - q = 0$ on that interval, and indeed, since (2.2.7) then simplifies to

$$\theta' = \cos 2\theta,$$

that $q = \Lambda$ throughout. But then $\Phi = \begin{pmatrix} \cosh & \sinh \\ \sinh & \cosh \end{pmatrix}$, and so the equation is unstable and (3.12.2) is not satisfied.

Thus (3.12.8) holds in both cases. Therefore there exists a constant c such that

$$I \bigg/ \int_0^a (v^T W v)^2 < c < \int_0^a (u^T W u)^2 \bigg/ I.$$

The choice $r = u^T W u - c v^T W v$ makes

$$\int_0^a u^T r W u > 0, \quad \int_0^a v^T r W v < 0,$$

and then the right-hand side of (3.12.6) is the sum of positive terms. Thus (3.12.7) is satisfied and the proof of the theorem is complete. $\square$

We conclude this section by mentioning an important situation which falls between those considered in sections 3.10-3.11 and in this section. Let the integer $M \ (\geq 1)$ be given and let $Q(M)$ denote the set of C^∞ potentials q for which Hill's equation has M consecutive double eigenvalues, i.e., M consecutive instability intervals vanish. Then we state the following theorem concerning the approximation of general potentials by those in $Q(M)$. We refer to [29] for the proof.

Theorem 3.12.2. *The set $Q(M)$ is dense in $\mathcal{F}$.*

3.13 Chapter notes

§3.2 The use of the Prüfer transformation for the spectral asymptotics of the Sturm-Liouville equation goes back to Hochstadt [85], [86]; see however the comment on these papers in the notes to §3.5 below. See also [48, chapter 4] and [127, §2.4].

§3.3 Theorems 3.3.1 and 3.3.2 appeared first in Eastham [48, Theorems 4.2.1 and 4.2.2], but the proof of the former given here is a new one.

For Theorem 3.3.3 and the other results of the same type, see Eastham [49], Guseinov and Yaslam [204], Karaca [105] and Ntinos [141].

§3.4 Theorem 3.4.1 is due to Titchmarsh [183, §21.5], where the proof uses the asymptotic form of the discriminant $D(\lambda)$ as $\lambda \to \infty$; see also [48, §4.4]. The proof given here is new in Brown and Eastham [22] and it produces an improved error term $O(m^{-2})$.

A less precise form of (3.4.3) which has only $o(m^{-1})$ in place of $m^{-1}|c_{2m+2}|$ goes back to Borg [17].

Coskun and Harris [30] give a different proof of (3.4.3) in which the O-term is replaced by $O(m^{-1}\eta^2(m)) + O(m^{-2}\eta(m))$ and $\eta(m)$ is of the order of the Fourier coefficients.

The step-function example (3.4.15) goes back to Kronig and Penney [120]; see also Titchmarsh [182] and [183, §21.5]. Our exposition here differs slightly from, and avoids an inaccuracy in, [182] and [183].

§3.5 The proof of Theorem 3.5.1 is new in Brown and Eastham [22], where it is also shown that this Prüfer method extends to produce the corresponding asymptotics for the periodic p-Laplacian. Theorem 3.5.1 itself is given in Marčenko and Ostrovskii [128, Theorem 4.2].

A less precise form of (3.5.1) which has only $o(m^{-r-1})$ in place of $m^{-1}|c_{2m+2}|$ goes back to Borg [17]. For an alternative method, see Levitan and Gasymov [126, pp. 58-60].

A proof of this form of (3.5.1) using the Prüfer method goes back to Hochstadt [85], [86]; see also [127, §2.4]. However, the details of the integration by parts in our proof of Theorem 3.5.1 are less simple than, and do not agree with, those in [85],[86].

Erovenko [57] uses the asymptotic form of $D(\lambda)$ to obtain a similar result to, but not quite in agreement with, Theorem 3.5.1.

Menken [133] gives an asymptotic formula corresponding to (3.5.1) (with $r = 3$) for the fourth-order equation $y^{(4)} + qy = \lambda y$.

§3.6 Lazutkin and Pankratova [125] give an estimate for l_n of the same type as in (iv) with a different, less explicit Fourier coefficient but an improved O-term.

There are results corresponding to (iv) for fourth-order equations in Erovenko [58], and for a restricted Dirac system in Simonyan [169].

3.13. Chapter notes

The proof of Theorem 3.6.1, which is new here, may be described as 'elementary' and it is a development of the method in Eastham [41]. An alternative proof can be deduced from the related work of Belokolos and Pershko [13].

In the context of inverse theory, Trubowitz [188, §3] shows that l_n is exponentially decreasing if and only if q is analytic, that is, there is also a converse relationship. See also Grigis [70] and Djakov and Mityagin [33], [35].

An earlier application, [43] and [48, Theorem 5.4.2], of Corollary 4.7.3 gives an estimate of l_n expressed explicitly in terms of the Fourier coefficients of q and valid for $n \geq n_0$ with a specific n_0. This is a generalization of the estimate $l_n \leq 2a^{-2} \left(\int_0^a q^2 \right)^2$, valid when q has mean-value zero and the mid-point of l_n is non-negative, due to Putnam [146] (see also [42, §6]).

Perhaps the first exponentially small estimates for l_n were given by Simonyan [168] for the equation $y''(x) + \lambda w(x) y(x) = 0$ in which x is taken as a complex variable with w analytic. The method uses concepts from the theory of differential equations in the complex domain.

For the Dirac system, Harris [76] uses his repeated diagonalization method [74, 75] to obtain a result corresponding to (3.6.2), which is that

$$l(\mu) \leq A|\mu|^{\frac{1}{2}} \exp(-B|\mu|) \qquad (|\mu| \to \infty)$$

with specified constants A and B, where $l(\mu)$ is the length of the instability interval centred on μ.

Inequality (3.6.33) holds for m sufficiently large, say $m \geq M(q)$. Hochstadt [88] introduced the problem of finding an explicit expression for $M(q)$ in terms of q, but his paper contained an error as pointed out in Brown and Eastham [23, §5.4]. Correct expressions for $M(q)$ were given in [23], where the method also covers the periodic p-Laplacian.

A Floquet-type analysis is also possible for a generalised Hill equation where the periodic potential q is replaced with a periodic arrangement of interface conditions (called *point interactions*). These may be of δ type (continuous solutions with jump in their derivatives), δ' type (jump in solutions with continuous derivative) or mixed type. In this set-up, the asymptotics of the lengths of stability and instability intervals differs from those of the standard Hill equation; for periodic δ, the lengths of instability intervals are asymptotically constant, while the lengths of the stability intervals grow linearly; conversely for periodic δ', the lengths of the instability intervals have linear asymptotic growth while the lengths of the stability intervals are asymptotically constant [61]. Periodic point interactions were studied further in [134], [123], [207] and, with regard to the vanishing of instability intervals, in [206].

§3.7 The nature of the dominant term in (3.7.2) was identified by Bell [11]; see also Harrell [73]. Our exposition here follows that of Avron and Simon [8] but with some simplification. Hochstadt [94] gives a different proof based on continued fraction expressions for eigenvalues [132, p.118].

Anahtarci and Djakov [4] have recently shown that the O-term in (3.7.2) is more precisely $-\frac{1}{4}c^2 n^{-3} + O(n^{-4})$.

Accurate numerical values of the λ_n and μ_n for the Mathieu equation go back to Ince [99], see also National Bureau of Standards, Tables relating to Mathieu functions, Columbia Univ. Press, 1951

Djakov and Mityagin [38] obtain a corresponding result to Theorem 3.7.1 for the two-term potential $q(x) = c\cos 2x - 2\alpha^2 \cos 4x$ (cf. section 2.8, Example 2). Grigis [70] gives a generic form of the dominant term in the asymptotics of l_n when q is a certain type of trigonometric polynomial.

Djakov and Mityagin [34] consider an analogue of (3.7.1) for the Dirac system (1.5.4) in which $p_1 = q = 0$ and $p_2 = 2c\cos 2x$ (see the notes on §§1.5 and 1.7). They obtain an asymptotic formula of the same nature as (3.7.2).

§**3.8** Hochstadt [90], Titchmarsh [183, §21.5], Goldberg [68, Appendix I and II].

§**3.9** This remarkable method is due to Hochstadt [90]; see also [48, Theorem 4.6.1], where errors in (4.6.11) and (4.6.12) are here corrected.

With λ as a complex variable, the proof that solutions of (3.9.1) have order $\frac{1}{2}$ is given in [127, §2.2] and [185, §1.5].

For the Hadamard Factorisation Theorem, see [181, §8.24].

Alternative proofs of Theorem 3.9.2 are the original one of Borg [17] and those of Hochstadt [87] and [92, pp. 307-308] and Ungar [189].

Korotyaev [113] has an interesting quantitative extension of Theorem 3.9.2 for the 1-periodic Hill equation with $p = w = 1$ and q square-integrable on the period interval, $\int_0^1 q = 0$. Denoting the $L^2([0,1])$ norm of q by $\|q\|$ and the ℓ^2 norm of the sequence of instability interval lengths (excluding the zero-th instability interval) by $\|\gamma\|$, he proves that $\|q\| \leq 2\|\gamma\|(1 + \|\gamma\|^{1/3})$. If, in addition, q is differentiable and q' is square-integrable on the period interval, a similar estimate in terms of the second moment of the squared sequence of instability interval lengths holds [115, Theorem 2.5]. Conversely, the norm of the sequence of instability interval lengths can be estimated in terms of the $L^2([0,1])$ norm of the potential q.

Analogous estimates hold for the periodic Dirac system with $q = 0$ [114, Theorem 1.1].

§**3.10** The exposition is based on the papers of Goldberg [68, 69] but with the simplification brought about by the choice (3.10.4) of λ early in the method.

Further to (3.10.7) and (3.10.12), S_n can be expressed explicitly in terms of S_j ($0 \leq j \leq n-1$), and the somewhat complicated formula is given in [2, (1.19)]. The S_n are related to a hierarchy of KdV equations (Weikard [197, §7]) and standard texts are Belokolos et al. [12] and Dubrovin et al. [40].

The case $N = 1$ appeared first in Hochstadt [90] where it was proved that q must be a Weierstrass elliptic function. Akhiezer [1] shows how to construct the elliptic function, given the instability interval. See also Trubowitz [188, pp. 325-326].

The first example of a potential with N gaps was given by Ince [100] (see also Erdélyi et al. [56, p.63ff]) where the potential is a multiple $N(N+1)$ of a . Weikard [197] gives a far-reaching development of this multiple in relation to the KdV hierarchy.

§3.11 Theorem 3.11.1 is due originally to Borg [17]. The proof here is based on, but develops, the method of Hochstadt [93]. In an earlier method of Hochstadt [92, pp. 306-307], it is not clear that the deduction on p. 307 can be made. See also Hochstadt [91].

Trubowitz [188, p. 325] (using methods which lie beyond the scope of this book) states a generalization of Corollary 3.11.2 which is that q has period a/r ($r \in \mathbb{N}$) if and only if I_n vanishes when $r \nmid n$.

§3.12 Theorem 3.12.1 is due to Simon [167] and Schmidt [158] for the Hill and Dirac cases, respectively.

We refer to [29] not only for the proof of Theorem 3.12.2 but also for other related references.

The technique of turning points of coexistence into small instability intervals by a periodic perturbation, as by (3.12.7) in the proof of Theorem 3.12.1, was used by Moser [137] to construct an example of a one-dimensional Schrödinger operator with limit-periodic potential for which the set of points of increase of the rotation number has the structure of a Cantor set. His construction is based on the observation that when an a-periodic equation is considered as $2a$-periodic, the zeros of the period-a discriminant become semi-periodic coexistence points of the period-$2a$ discriminant. Thus they subsequently open up into an instability interval when a suitable $2a$-periodic perturbation is added, splitting a previous stability interval into two components, whose lengths can be controlled by the estimate of Theorem 2.4.4. A procedure of alternate period doubling and judicious perturbation results in a limit-periodic problem where all stability intervals have disintegrated and the instability intervals are dense on the real line.

Chapter 4

Spectra

4.1 Introduction

In section 1.8 we studied the eigenvalues of the periodic, semi-periodic and twisted boundary-value problem on a period interval, and similarly those of a boundary-value problem with separated boundary conditions in section 2.3. We begin the present chapter with the observation that the eigenvalues and eigenfunctions of these boundary-value problems give rise to a generalisation of Fourier series convergent in the sense of a suitable Hilbert space. Thence we proceed to study the spectral properties of the periodic equation on the real line and on a half-line with a boundary condition of the separated type.

In section 4.3 we perform the limiting process of sending one of the end-points of a bounded interval to infinity, finding a limit for the spectral functions governing the generalised Fourier transforms of the boundary-value problems on the interval. This limit is a spectral function for the half-line problem. The associated self-adjoint operator, constructed in section 4.4, will in general have intervals of continuous spectrum in addition to eigenvalues. In the special case of the periodic equation on the half-line, we show in section 4.5 that the spectrum in fact consists of purely absolutely continuous bands coinciding with the closure of the stability set S, with at most a single eigenvalue in each instability interval.

The full-line case has the technical complication that the spectral function will be matrix-valued. However, the spectral structure of the periodic equation on the whole real line is even simpler than on the half-line, comprising the absolutely continuous bands only.

We conclude the chapter with an extension of the oscillation method of counting eigenvalues, introduced in section 2.3 for the separated boundary-value problem on the period interval, to the half-line case. This will prove a valuable tool for the study of the perturbed periodic equation in Chapter 5.

4.2 Regular boundary-value problems

We again consider the boundary-value problem comprising (1.5.1) on the interval $[0, a]$ with either the separated boundary conditions

$$u_1(0)\cos\alpha - u_2(0)\sin\alpha = 0, \quad u_1(a)\cos\beta - u_2(a)\sin\beta = 0 \qquad (4.2.1)$$

as in (2.3.2), or the periodic, semi-periodic and ω-twisted condition

$$u(a) = \omega u(0) \qquad (4.2.2)$$

as in (1.8.1). Our purpose in this section is to put the previous spectral theory into the context of compact operators in the Hilbert space

$$L^2([0,a];W) := \{f : [0,a] \to \mathbb{C}^2 \mid f \text{ measurable}, \int_0^a f^T W \overline{f} < \infty\}/\sim,$$

where the equivalence relation $\sim$ is defined as

$$f \sim g :\Leftrightarrow \int_0^a (f-g)^T W \overline{(f-g)} = 0$$

and $\{\ldots\}/\sim$ is the set of equivalence classes. The inner product is defined as

$$(f,g) := \int_0^a f^T W \overline{g} \qquad (f, g \in L^2([0,a];W)),$$

and we denote the corresponding norm by $\|\cdot\|$.

The general assumptions on B and W are as in section 1.5, but our attention is confined to the interval $[0, a]$, and a-periodicity is not a feature in this section.

We recall that, in the case of the Sturm-Liouville system (1.5.2),

$$W = \begin{pmatrix} w & 0 \\ 0 & 0 \end{pmatrix},$$

and so $L^2([0,a];W) = L^2([0,a];w)$, the standard L^2 space of complex-valued functions with weight w. For the Dirac system (1.5.4), $W = I$, and so $L^2([0,a];W) = L^2(0,a)^2$, the space of $\mathbb{C}^2$-valued, square-integrable functions.

We begin with the inhomogeneous equation

$$u' = J(B + \lambda W)u + JWf, \qquad (4.2.3)$$

where $f : [0, a] \to \mathbb{C}^2$ is a given function, and we seek a solution U which satisfies either (4.2.1) or (4.2.2). By the variation of constants formulae (1.2.10) and (1.2.13), the general solution U of (4.2.3) is

$$U(x) = \Psi(x)c + u(x)\int_0^x \frac{v^T W f}{\det \Psi} + v(x)\int_x^a \frac{u^T W f}{\det \Psi}, \qquad (4.2.4)$$

4.2. Regular boundary-value problems

where c is a constant, u and v are linearly independent solutions of the homogeneous equation (1.5.1), and $\Psi = (u, v)$. If λ is not an eigenvalue for (1.5.1) and the relevant boundary condition, the solution U is unique because, otherwise, the difference of two distinct solutions would be an eigenfunction corresponding to λ. The resulting unique relationship between U and f defines via (4.2.4) an integral operator which is known as the *resolvent operator* for the boundary-value problem comprising (1.5.1) and the relevant boundary condition. The resolvent operator depends on $\lambda \in \mathbb{C}$ and it is defined for λ not an eigenvalue.

We construct the resolvent operator first for the case of separated boundary conditions (4.2.1). Let u and v be the solutions of (1.5.1) with the initial values

$$u(a) = \begin{pmatrix} \sin \beta \\ \cos \beta \end{pmatrix}, \qquad v(0) = \begin{pmatrix} \sin \alpha \\ \cos \alpha \end{pmatrix}.$$

Then u and v are linearly independent if λ is not an eigenvalue. Also, in (4.2.4), the choice $c = 0$ gives the unique U which satisfies (4.2.1). The integral kernel of the resolvent operator is the *Green's function* $\mathcal{G}_\lambda(x, t)$ defined by

$$\mathcal{G}_\lambda(x, t) := \begin{cases} u(x)v(t)^T / \det \Psi(t) & \text{if } t < x, \\ v(x)u(t)^T / \det \Psi(t) & \text{if } t > x. \end{cases}$$

It is an essentially bounded 2×2 matrix-valued function with the symmetry property that $\mathcal{G}_\lambda(x, t) = \mathcal{G}_\lambda(t, x)^T$. For $\lambda \in \mathbb{R}$, $\mathcal{G}_\lambda$ has real entries. Now (4.2.4) can be written as $U(x) = \int_0^a \mathcal{G}_\lambda(x, t) W(t) f(t) dt$ and, in the next theorem, we express this relation and its properties in terms of the resolvent operator.

Theorem 4.2.1. *Suppose that $\lambda \in \mathbb{C}$ is not an eigenvalue of the boundary-value problem (1.5.1), (4.2.1). Then the resolvent operator $\mathcal{R}_\lambda$ defined by*

$$(\mathcal{R}_\lambda f)(x) := \int_0^a \mathcal{G}_\lambda(x, t) W(t) f(t) \, dt \qquad (x \in [0, a]; f \in L^2([0, a]; W))$$

is a bounded linear operator in $L^2([0, a]; W)$. In fact, it is a compact operator. If $\lambda \in \mathbb{R}$, then $\mathcal{R}_\lambda$ is symmetric.

Proof. We first show that, for each $\lambda \in \mathbb{C}$ and $f \in L^2([0, a]; W)$, $\mathcal{R}_\lambda f$ is a continuous function on $[0, a]$. Indeed, for $0 \leq x < z \leq a$,

$$|(\mathcal{R}_\lambda f)(x) - (\mathcal{R}_\lambda f)(z)| \leq \int_0^x |\mathcal{G}_\lambda(x, t) - \mathcal{G}_\lambda(z, t)| \, |W(t) f(t)| \, dt$$

$$+ \int_x^z (|\mathcal{G}_\lambda(x, t)| + |\mathcal{G}_\lambda(z, t)|) \, |W(t) f(t)| \, dt + \int_z^a |\mathcal{G}_\lambda(x, t) - \mathcal{G}_\lambda(z, t)| \, |W(t) f(t)| \, dt$$

$$\leq \max \left\{ \sup_{t < x} \left| \frac{(u(x) - u(z))v(t)^T}{\det \Psi(t)} \right|, \sup_{t > z} \left| \frac{(v(x) - v(z))u(t)^T}{\det \Psi(t)} \right| \right\} \left(\int_0^a |W| \right)^{1/2} \|f\|$$

$$+ 2\|\mathcal{G}_\lambda\|_\infty \left(\int_x^z |W| \right)^{1/2} \|f\|$$

$$\to 0 \qquad (|x - z| \to 0), \tag{4.2.5}$$

since u and v are uniformly continuous and $|W|$ is locally integrable. Similarly, for any $x \in [0, a]$,

$$|(\mathcal{R}_\lambda f)(x)| \leq \|\mathcal{G}_\lambda\|_\infty \left(\int_0^a |W|\right)^{1/2} \|f\|, \qquad (4.2.6)$$

and so

$$\|\mathcal{R}_\lambda f\| \leq \|\mathcal{G}_\lambda\|_\infty \left(\int_0^a |W|\right) \|f\|,$$

showing that $\mathcal{R}_\lambda$ is a bounded operator.

If $(f_n)_{n \in \mathbb{N}}$ is a bounded sequence in $L^2([0,a];W)$, then the sequence of functions $(\mathcal{R}_\lambda f_n)$ is uniformly bounded by (4.2.6) and equicontinuous by (4.2.5). Hence, by the Arzelà-Ascoli Theorem, it has a uniformly convergent subsequence, which is also convergent in $L^2([0,a];W)$. Thus the operator $\mathcal{R}_\lambda$ is compact.

If $\lambda \in \mathbb{R}$, then

$$(\mathcal{R}_\lambda f, g) = \int_0^a \int_0^a f(t)^T W(t) \mathcal{G}_\lambda(x,t)^T W(x) \overline{g(x)} \, dt \, dx$$

$$= \int_0^a \int_0^a f(t)^T W(t) \overline{\mathcal{G}_\lambda(t,x)} W(x) \overline{g(x)} \, dx \, dt = (f, \mathcal{R}_\lambda g)$$

and thus $\mathcal{R}_\lambda$ is symmetric. □

In the case of the Sturm-Liouville system, only the components f_1 and $(\mathcal{G}_\lambda)_{11}$ are significant due to the structure of the matrix W; hence $(\mathcal{G}_\lambda)_{11}$ is usually called the Green's Function for the Sturm-Liouville equation. This function is expressed in terms of the function values of the solutions in a fundamental system of the Sturm-Liouville equation only, without reference to their quasi-derivatives.

In a similar way, we can construct the resolvent operator for the case of the boundary condition (4.2.2). Suppose again that $\lambda \in \mathbb{C}$ is not an eigenvalue for the boundary-value problem (1.5.1), (4.2.2). We first choose $\alpha = \beta \in (0, \pi)$ such that λ is also not an eigenvalue for the boundary-value problem (1.5.1), (4.2.1). For the Sturm-Liouville and Dirac systems this can always be achieved by suitable choice of α in view of Theorem 2.4.5 with $\tau = 0$.

With this value of $\alpha(= \beta)$, let u, v and Ψ be as above. Then $u(a) = v(0)$. Also, u and v are linearly independent, and there is no $c \in \mathbb{C}^2 \setminus \{0\}$ such that $(\Psi(a) - \omega \Psi(0))c = 0$, as otherwise Ψc would be an eigenfunction corresponding to λ. Therefore $\Psi(a) - \omega \Psi(0)$ is regular, and the choice

$$c := (\Psi(a) - \omega \Psi(0))^{-1} u(a) \int_0^a \frac{(\omega u - v)^T W f}{\det \Psi}$$

in (4.2.4) gives a solution U for the inhomogeneous equation such that $U(a) = \omega U(0)$.

4.2. Regular boundary-value problems

Taking $\mathcal{G}_\lambda$ to be the Green's function for the separated boundary-value problem with $\alpha = \beta$ and abbreviating

$$\mathcal{L}_\lambda(x) = \Psi(x)(\Psi(a) - \omega\Psi(0))^{-1}\Psi(a)\begin{pmatrix}1\\0\end{pmatrix},$$

$$\mathcal{H}_\lambda(x) = (\omega u(x) - v(x))^T / \det \Psi(x),$$

considerations analogous to the proof of Theorem 4.2.1 lead to the following statement.

Theorem 4.2.2. *Suppose that $\lambda \in \mathbb{C}$ is not an eigenvalue of the boundary-value problem* (1.5.1), (4.2.2). *Then the resolvent operator $\mathcal{R}_\lambda$ defined by*

$$(\mathcal{R}_\lambda f)(x) := \int_0^a \mathcal{G}_\lambda(x,t) W(t) f(t)\, dt + \mathcal{L}_\lambda(x) \int_0^a \mathcal{H}_\lambda(t) W(t) f(t)\, dt,$$

$$(x \in [0,a]; f \in L^2([0,a];W))$$

is a bounded, compact linear operator in $L^2([0,a]; W)$. If $\lambda \in \mathbb{R}$, then $\mathcal{R}_\lambda$ is symmetric.

The statement about the symmetry can be shown by recourse to the differential equation. If $f, g \in L^2([0,a]; W)$ and $U := \mathcal{R}_\lambda f$, $V := \mathcal{R}_\lambda g$, then

$$(\mathcal{R}_\lambda f, g) = -\int_0^a U^T(J\overline{V}' + (B+\lambda W)\overline{V})$$

$$= -\int_0^a (JU' + (B+\lambda W)U)^T \overline{V} = (f, \mathcal{R}_\lambda g)$$

by an integration by parts; the boundary terms cancel because U and V satisfy (4.2.2) with $|\omega| = 1$.

Theorems 4.2.1 and 4.2.2 show that the resolvent operators for the boundary conditions (4.2.1) and (4.2.2) satisfy the hypotheses of the Hilbert-Schmidt theorem, which states that a compact symmetric linear operator in a Hilbert space has a sequence of real eigenvalues which converges to 0, and that the corresponding eigenfunctions can be chosen to form an orthonormal basis of the Hilbert space [149]. We can relate these general properties of eigenfunctions to the boundary-value problems with (4.2.1) and (4.2.2) as follows. Suppose that μ is not an eigenvalue of the boundary-value problem, and let λ be an eigenvalue with eigenfunction u. Then, by (1.5.1),

$$u' = J(B + \mu W)u + (\lambda - \mu)JWu.$$

Hence
$$u = (\lambda - \mu)\mathcal{R}_\mu u, \qquad (4.2.7)$$

and so u is an eigenfunction of $\mathcal{R}_\mu$ with eigenvalue $(\lambda - \mu)^{-1}$. Thus any eigenfunction u is also an eigenfunction of the corresponding $\mathcal{R}_\mu$.

We therefore obtain the following theorem in which, as before, $\mathcal{J} = \mathbb{N}_0$ for the Sturm-Liouville case and $\mathcal{J} = \mathbb{Z}$ for the Dirac case.

Theorem 4.2.3. *The boundary-value problem (1.5.1), with boundary conditions either (4.2.1) or (4.2.2), has a set of eigenfunctions $\{u_j \mid j \in \mathcal{J}\}$ which form an orthonormal basis of $L^2([0,a]; W)$.*

For any $f \in L^2([0,a]; W)$, the Parseval formula

$$\|f\|^2 = \sum_{j \in \mathcal{J}} |(f, u_j)|^2 \tag{4.2.8}$$

and the expansion formula

$$f = \sum_{j \in \mathcal{J}} (f, u_j) u_j \tag{4.2.9}$$

are valid.

In the simplest case of a Sturm-Liouville equation with $p = w = 1$ and $q = 0$, the Dirichlet eigenfunctions are $\sin(j\pi x/a)$, $j \in \mathbb{N}$, and so after suitable normalisation (4.2.9) describes the expansion of a function f in a Fourier sine series. In addition, the periodic boundary conditions produce the eigenfunctions 1 and $\cos(2j\pi x/a)$, $\sin(2j\pi x/a)$. Then (4.2.9) becomes the full Fourier expansion of f.

Corollary 4.2.4. *Let $\mu \in \mathbb{R}$ and let $\hat{\lambda}$ be the eigenvalue for the boundary-value problem (1.5.1), with boundary conditions either (4.2.1) or (4.2.2), which is closest to μ. Then*

$$\|\mathcal{R}_\mu\| = \frac{1}{|\hat{\lambda} - \mu|}.$$

Proof. Let $\hat{u}$ be an eigenfunction for $\hat{\lambda}$. Then, as in (4.2.7),

$$\|\mathcal{R}_\mu \hat{u}\| = \left\| \frac{1}{\hat{\lambda} - \mu} \hat{u} \right\| = \frac{\|\hat{u}\|}{|\hat{\lambda} - \mu|},$$

and so $\|\mathcal{R}_\mu\| \geq 1/|\hat{\lambda} - \mu|$.

Conversely, let $f \in L^2([0,a]; W)$. Since $\mathcal{R}_\mu$ is linear and bounded, we have from (4.2.9)

$$\mathcal{R}_\mu f = \sum_{j \in \mathcal{J}} (f, u_j) \mathcal{R}_\mu u_j = \sum_{j \in \mathcal{J}} (f, u_j) \frac{1}{\lambda_j - \mu} u_j.$$

Hence by (4.2.8)

$$\|\mathcal{R}_\mu f\|^2 = \sum_{j \in \mathcal{J}} |(f, u_j)|^2 / (\lambda_j - \mu)^2 \leq \frac{1}{|\hat{\lambda} - \mu|^2} \|f\|^2.$$

This proves the opposite inequality $\|\mathcal{R}_\mu\| \leq 1/|\hat{\lambda} - \mu|$. $\square$

4.2. Regular boundary-value problems

This leads to the following stability result for the eigenvalues under perturbations.

Theorem 4.2.5. *Let the boundary-value problem* (1.5.1), *with boundary conditions either* (4.2.1) *or* (4.2.2), *have no eigenvalues in an interval* $(\mu - \delta, \mu + \delta)$. *Let* $q_1 : [0, b] \to \mathbb{R}$ *be such that*

$$\|q_1\|_\infty := \operatorname{ess\ sup}_{x \in [0,a]} |q_1(x)| < \delta.$$

Then the perturbed boundary-value problem

$$u' = J(B + q_1 W + \lambda W) u$$

with the same boundary conditions has no eigenvalues in the interval $(\mu - \delta + \|q_1\|_\infty, \mu + \delta - \|q_1\|_\infty)$.

Proof. We begin by finding a relation between the resolvent operators for the original and the perturbed problems, denoted by $\mathcal{R}_\mu$ and $\mathcal{R}_{1,\mu}$ respectively. Let $f \in L^2([0,a];W)$ and let $U = \mathcal{R}_\mu f$, $V = \mathcal{R}_{1,\mu} f$. Then, by (4.2.3) and its perturbed analogue,

$$(U - V)' = J(B + q_1 W + \lambda W)(U - V) - JW q_1 U,$$

and so $U - V = -\mathcal{R}_{1,\mu} q_1 U$. Solving for V, we find that

$$\mathcal{R}_{1,\mu} f = \mathcal{R}_\mu f + \mathcal{R}_{1,\mu} q_1 \mathcal{R}_\mu f$$

and hence

$$\|\mathcal{R}_{1,\mu}\| = \sup_f \frac{\|\mathcal{R}_{1,\mu} f\|}{\|f\|}$$

$$\leq \sup_f (1 + \|\mathcal{R}_{1,\mu}\| \|q_1\|_\infty) \frac{\|\mathcal{R}_\mu f\|}{\|f\|} = (1 + \|\mathcal{R}_{1,\mu}\| \|q_1\|_\infty) \|\mathcal{R}_\mu\|.$$

Solving for $\|\mathcal{R}_{1,\mu}\|$ and noting that $\|\mathcal{R}_\mu\| \leq 1/\delta$ by Corollary 4.2.4, we obtain

$$\|\mathcal{R}_{1,\mu}\| \leq \frac{\|\mathcal{R}_\mu\|}{1 - \|\mathcal{R}_\mu\| \|q_1\|_\infty} \leq \frac{1}{\delta - \|q_1\|_\infty},$$

and the theorem follows by Corollary 4.2.4. □

The Parseval and expansion formulae (4.2.8) and (4.2.9) can be recast in an equivalent format in terms of Stieltjes integrals, and it is in this form that the formulae will be developed later in the chapter when the x-interval is $[0, \infty)$ with a boundary condition at $x = 0$ only. The relevant boundary conditions for this format are the separated ones (4.2.1). The orthonormal eigenfunctions u_j can be obtained by normalisation of the solution $v(\cdot, \lambda)$ of (1.5.1) with the initial data

$$v(0, \lambda) = \begin{pmatrix} \sin \alpha \\ \cos \alpha \end{pmatrix}. \tag{4.2.10}$$

Then $v(\cdot, \lambda)$ satisfies the boundary condition at 0 and it will also satisfy the boundary condition at a if and only if λ is an eigenvalue. Thus

$$u_j(x) = v(x, \lambda_j) \Big/ \sqrt{\int_0^a v(x, \lambda_j)^T W(x) \overline{v(x, \lambda_j)} \, dx},$$

and we can write (4.2.8) as

$$\|f\|^2 = \sum_{j \in \mathcal{J}} |g(\lambda_j)|^2 \Big/ \int_0^a v(x, \lambda_j)^T W(x) \overline{v(x, \lambda_j)} \, dx, \qquad (4.2.11)$$

where

$$g(\lambda) := \int_0^a f(x)^T W(x) \overline{v(x, \lambda)} \, dx \qquad (\lambda \in \mathbb{C}) \qquad (4.2.12)$$

is a *generalised Fourier transform* of f. From (4.2.11), the Parseval formula can now be written equivalently in the form of a Stieltjes integral,

$$\int_0^a f(x)^T W(x) \overline{f(x)} \, dx = \int_{\mathbb{R}} |g(\lambda)|^2 \, d\sigma_{a,\beta}(\lambda), \qquad (4.2.13)$$

with the *spectral function* $\sigma_{a,\beta}$ defined for real λ by

$$\sigma_{a,\beta}(\lambda) = \begin{cases} \sum_{\lambda_j \in (0,\lambda]} \left(\int_0^a v(x, \lambda_j)^T W(x) \overline{v(x, \lambda_j)} \, dx \right)^{-1} & \text{if } \lambda \geq 0, \\ -\sum_{\lambda_j \in (\lambda, 0]} \left(\int_0^a v(x, \lambda_j)^T W(x) \overline{v(x, \lambda_j)} \, dx \right)^{-1} & \text{if } \lambda < 0. \end{cases} \qquad (4.2.14)$$

Similarly, the expansion formula (4.2.9) takes the form

$$f(x) = \int_{-\infty}^{\infty} g(\lambda) \, v(x, \lambda) \, d\sigma_{a,\beta}(\lambda), \qquad (4.2.15)$$

where the equality is understood in the sense of a limit in $L^2([0, a]; W)$.

4.3 The spectral function for the half-line problem

A formal argument going back to Fourier himself considers the Fourier sine series (and similarly the cosine series) of a function f on a large interval $[0, b]$, in terms of $\sin(j\pi x/b)$ ($j \in \mathbb{N}$). Then, when $b \to \infty$, the Fourier sine integral of f appears as a result of the limiting process. A similar, but rigorous, procedure applies to (4.2.13) and (4.2.15), with a replaced by b, for developing the spectral theory of (1.5.1) on the half-line $[0, \infty)$ with a single boundary condition

$$u_1(0) \cos \alpha - u_2(0) \sin \alpha = 0 \qquad (4.3.1)$$

4.3. The spectral function for the half-line problem

at the finite end-point. The starting point for this development is therefore (1.5.1) considered to hold on $[0, b]$ with, as in (4.2.1), separated boundary conditions

$$u_1(0) \cos \alpha - u_2(0) \sin \alpha = 0, \quad u_1(b) \cos \beta - u_2(b) \sin \beta = 0. \tag{4.3.2}$$

In the limiting process $b \to \infty$, a-periodicity of B and W plays no part and will not be assumed. In the next section however, a-periodicity will feature again as we move on to the specialised consequences of the general theory for periodic systems.

We begin with a simple lemma.

Lemma 4.3.1. *Let* $U, V : [0, b] \to \mathbb{C}^2$ *be solutions of (1.5.1) with spectral parameters* $\lambda \in \mathbb{C}$, $\mu \in \mathbb{C}$, *respectively. Then*

$$(U^T J \overline{V}(b) - (U^T J \overline{V})(0) = (\lambda - \overline{\mu}) \int_0^b U^T W \overline{V}.$$

Proof. From (1.5.1)

$$\int_0^a (\lambda U^T W V - U^T W \overline{\mu V})$$

$$= -\int_0^b \left(JU' + BU \right)^T - U^T \overline{(JV' + BV)} \right) = \int_0^b (U^T J \overline{V})'$$

since $J^T = -J$ and the lemma follows. $\square$

With a fixed boundary condition (4.3.1) at 0, we consider the fundamental system $u(\cdot, \lambda)$ and $v(\cdot, \lambda)$ of (1.5.1) such that

$$u(0, \lambda) = \begin{pmatrix} \cos \alpha \\ -\sin \alpha \end{pmatrix}, \quad v(0, \lambda) = \begin{pmatrix} \sin \alpha \\ \cos \alpha \end{pmatrix} \tag{4.3.3}$$

for each $\lambda \in \mathbb{C}$. In particular, v satisfies the first boundary condition in (4.3.2). Now, given β, there is a non-trivial solution of (1.5.1) satisfying the boundary condition at b in (4.3.2), and it can be expressed as a linear combination of u and v. The coefficient of u in this linear combination is non-zero if $\lambda \in \mathbb{C} \setminus \mathbb{R}$ because, otherwise, v would be an eigenfunction for the separated boundary conditions (4.3.2) with a non-real eigenvalue, and no such eigenvalues exist. Hence, on multiplication by a suitable constant, the coefficient of u can be made equal to 1. We therefore arrive at our first fundamental observation that, for non-real λ, there exists a unique $m_{b,\beta}(\lambda) \in \mathbb{C}$ such that

$$u(\cdot, \lambda) + m_{b,\beta}(\lambda) v(\cdot, \lambda) \tag{4.3.4}$$

satisfies the boundary condition at b in (4.3.2). This solution of (1.5.1) provides the basis for the spectral theory of (1.5.1) on $[0, \infty)$, and the following lemma is the first main step. In the lemma, we continue to take $\lambda \in \mathbb{C} \setminus \mathbb{R}$ and we give the proof for $\operatorname{Im} \lambda > 0$, with $\operatorname{Im} \lambda < 0$ being similar.

Lemma 4.3.2. (a) $m_{b,\beta}(\lambda)$ lies on the complex circle of radius

$$(2|\operatorname{Im}\lambda| \int_0^b v^T W \bar{v})^{-1} \qquad (4.3.5)$$

and centre $-(u^T J\bar{v})(b)/(v^T J\bar{v})(b)$. If $m \in \mathbb{C}$ is on this circle, then

$$\int_0^b (u+mv)^T W \overline{u+mv} = \frac{\operatorname{Im} m}{\operatorname{Im} \lambda}; \qquad (4.3.6)$$

if $m \in \mathbb{C}$ is inside the circle, then

$$\int_0^b (u+mv)^T W \overline{u+mv} < \frac{\operatorname{Im} m}{\operatorname{Im} \lambda}. \qquad (4.3.7)$$

(b) If $0 < b_1 < b_2$, then the circle for b_2 lies inside the circle for b_1.

Proof. (a) Let $m = m_{b,\beta}(\lambda)$. Then, since $u+mv$ satisfies the boundary condition at b,

$$0 = ((u+mv)^T J \overline{u+mv})(b) = X - m\overline{Y} + \overline{m}Y + |m|^2 Z, \qquad (4.3.8)$$

where we abbreviate

$$X := (u^T J\bar{u})(b), \qquad Y := (u^T J\bar{v})(b), \qquad Z := (v^T J\bar{v})(b).$$

Since $(v^T J\bar{v})(0) = 0$ by (4.3.3), Lemma 4.3.1 with $U = V = v$ gives

$$Z = 2i \operatorname{Im} \lambda \int_0^b v^T W \bar{v},$$

which is non-zero and purely imaginary. Writing $m = z - Y/Z$, we find from (4.3.8) that

$$Z|z|^2 + \frac{XZ + |Y|^2}{Z} = 0. \qquad (4.3.9)$$

By a straightforward, if slightly tedious, calculation

$$XZ + |Y|^2 = |u^T Jv|^2(b) = 1, \qquad (4.3.10)$$

since $u^T Jv = 1$ is the Wronskian of the fundamental system. This gives the equation for a circle,

$$|z|^2 = |Z|^{-2} = \left| 2 \operatorname{Im} \lambda \int_0^b v^T W \bar{v} \right|^{-2}.$$

Now let $m \in \mathbb{C}$; if m is on the circle, then by (4.3.8), (4.3.9) and (4.3.10)

$$\frac{1}{2i}((u+mv)^T J \overline{u+mv})(b) = \left(\operatorname{Im} \lambda \int_0^b v^T W \bar{v} \right) |z|^2 + \frac{1}{2iZ} = 0. \qquad (4.3.11)$$

4.3. The spectral function for the half-line problem

Since the bracketed term on the right-hand side is positive, and $|z|$ is the distance of m from the circle centre, we find that m is inside the circle if and only if

$$\frac{1}{2i}((u+mv)^T J \overline{u+mv})(b) < 0. \tag{4.3.12}$$

Since $((u+mv)^T J \overline{u+mv})(0) = -2i \operatorname{Im} m$ in view of the initial data (4.3.3), Lemma 4.3.1 gives

$$\frac{1}{2i}((u+mv)^T J \overline{u+mv})(b) = \operatorname{Im} \lambda \int_0^b (u+mv)^T W \overline{u+mv} - \operatorname{Im} m. \tag{4.3.13}$$

Combined with (4.3.11) and (4.3.12), this completes the proof of part (a).

(b) Let m be a point on the circle for b_2. Then by (4.3.11) and (4.3.13)

$$\operatorname{Im} \lambda \int_0^{b_1} (u+mv)^T W \overline{u+mv} - \operatorname{Im} m < \operatorname{Im} \lambda \int_0^{b_2} (u+mv)^T W \overline{u+mv} - \operatorname{Im} m = 0,$$

so the statement follows using (4.3.7). □

Lemma 4.3.2 (b) shows that, for each fixed $\lambda \in \mathbb{C}\backslash\mathbb{R}$, the circles for b form a decreasing nested family as b increases. Hence they either shrink to a point (*limit-point case*) or approach a limit circle (*limit-circle case*) as $b \to \infty$. By (4.3.5), the limit-point case occurs if $v(\cdot, \lambda) \notin L^2([0,\infty); W)$ and the limit-circle case if $v(\cdot, \lambda) \in L^2([0,\infty); W)$. In either case, if m is the limit point or a point on the limit circle, then (4.3.7) holds for all $b > 0$, showing that (1.5.1) has a solution

$$u(\cdot, \lambda) + m\, v(\cdot, \lambda) \in L^2([0,\infty); W) \tag{4.3.14}$$

for all λ with $\operatorname{Im} \lambda \neq 0$.

It follows from (4.3.14) that, in the limit-circle case, all solutions of (1.5.1) lie in the space $L^2([0,\infty); W)$ when $\lambda \in \mathbb{C}\backslash\mathbb{R}$ because v and $u+mv$ form a fundamental system with this property. This observation, together with the next proposition, shows that the limit-point, limit-circle classification of (1.5.1) does not depend on the value of λ.

Proposition 4.3.3. *Let all solutions of (1.5.1) lie in $L^2([0,\infty); W)$ for any one value of $\lambda \in \mathbb{C}$. Then the same is true for any other value $\mu \in \mathbb{C}$.*

Proof. Let Φ be the canonical fundamental matrix for (1.5.1) with spectral parameter λ. Let $\mu \in \mathbb{C}$, and let v be any solution of (1.5.1) with spectral parameter μ. Then

$$v' = J(B + \lambda W)v + J(\mu - \lambda)Wv,$$

and the variation of constants formulae (1.2.10) and (1.2.11) show that

$$v(x) = \Phi(x)c + \Phi(x) \int_0^x \Phi^{-1} J(\mu - \lambda) W v \tag{4.3.15}$$

with some constant $c \in \mathbb{C}^2$. With regard to the first term in (4.3.15), we note that

$$C := \int_0^\infty |\Phi^T W \overline{\Phi}| < \infty$$

by hypothesis. We set $V(x) = \int_0^x v^T W \overline{v}$ and we have to show that $V(x)$ is bounded for all $x > 0$. Then, by (1.2.15) and the Cauchy-Schwarz inequality,

$$\left| \int_0^x \Phi^{-1} JWv \right| \leq \int_0^x |\Phi^T W^{\frac{1}{2}}| |W^{\frac{1}{2}} v|$$

$$\leq \sqrt{\int_0^x |\Phi^T W \overline{\Phi}|} \sqrt{V(x)} \leq \sqrt{CV(x)},$$

and hence (4.3.15) gives

$$V(x) \leq 2 \int_0^x (\Phi c)^T W \overline{\Phi c}$$
$$+ 2|\mu - \lambda|^2 \int_0^x \left(\int_0^t \Phi^{-1} JWv \right)^T \Phi(t)^T W(t) \overline{\Phi(t)} \overline{\int_0^t \Phi^{-1} JWv} \, dt$$
$$\leq 2C |c|^2 + 2|\mu - \lambda|^2 \int_0^x |\Phi(t)^T W(t) \overline{\Phi(t)}| \left| \int_0^t \Phi^{-1} JWv \right|^2 dt$$
$$\leq 2C |c|^2 + 2|\mu - \lambda|^2 C \int_0^x |\Phi^T W \overline{\Phi}| V.$$

This is a Gronwall inequality for V, and it follows that

$$V(x) \leq 2C |c|^2 \exp\left(2|\mu - \lambda|^2 C \int_0^x |\Phi^T W \overline{\Phi}|\right) \leq 2C |c|^2 \exp(2|\mu - \lambda|^2 C^2) < \infty$$

for all $x \geq 0$, as required. □

Thus we have *Weyl's Alternative* which states that, independently of $\lambda \in \mathbb{C}\backslash\mathbb{R}$, (1.5.1) is in either the limit-point case or the limit-circle case. In the limit-point case, the limit-point m is unique for each $\lambda \in \mathbb{C}\backslash\mathbb{R}$, and the resulting function of λ is called the *Titchmarsh-Weyl m-function*.

As observed at the end of section 1.4, the a-periodic system has a Floquet solution unbounded at ∞ if the spectral parameter is non-real; this implies the following.

Proposition 4.3.4. *The system (1.5.1) with a-periodic B and W is in the limit-point case.*

Our aim is now to show that the Parseval formula (4.2.13) extends to the half-line problem, with a spectral function which arises as the limit of the spectral functions $\sigma_{b,\beta}$ of the boundary-value problems on $[0, b]$ with boundary condition β at b, as $b \to \infty$.

4.3. The spectral function for the half-line problem

Theorem 4.3.5. *Assume that the boundary-value problem* (1.5.1), (4.3.1) *on* $[0,\infty)$ *is in the limit-point case. Then there exists a non-decreasing function* $\sigma : \mathbb{R} \to \mathbb{R}$ *such that*

(a)
$$\sigma(\lambda'') - \sigma(\lambda') = \lim_{b \to \infty} (\sigma_{b,\beta}(\lambda'') - \sigma_{b,\beta}(\lambda'))$$

for all λ' *and* $\lambda'' \in \mathbb{R}$ *where* σ *is continuous. The boundary condition parameter* β *may depend on* b *in an arbitrary way.*

(b)
$$\sigma(\lambda'') - \sigma(\lambda') = \lim_{\epsilon \to 0} \frac{1}{\pi} \int_{\lambda'}^{\lambda''} \operatorname{Im} m(t + i\epsilon) \, dt \qquad (4.3.16)$$

for all λ' *and* $\lambda'' \in \mathbb{R}$ *where* σ *is continuous.*

(c) *There exists a constant* $c \in \mathbb{R}$ *such that*

$$\operatorname{Im} m(\lambda) = \operatorname{Im} \lambda \int_{-\infty}^{\infty} \frac{d\sigma(\nu)}{|\nu - \lambda|^2} + c \qquad (\lambda \in \mathbb{C} \setminus \mathbb{R}). \qquad (4.3.17)$$

Proof. We show first that, for any fixed $\Lambda \in \mathbb{R}$, $\sigma_{b,\beta}(\Lambda)$ is bounded independently of b and β. This will allow us to apply the Helly Selection Theorem concerning convergent subsequences. Let u and v be the solutions of (1.5.1) defined by the initial data (4.3.3). For the value l ($\in \mathbb{C}\setminus\mathbb{R}$) of the spectral parameter, we have the solution

$$\psi_{b,\beta}(\cdot) := u(\cdot, l) + m_{b,\beta}(l) \, v(\cdot, l) \qquad (4.3.18)$$

which, as in (4.3.4), satisfies the boundary condition at b in (4.3.2). For $\lambda \in \mathbb{R}$, $v(\cdot, \lambda)$ is real-valued and, by Lemma 4.3.1,

$$\psi_{b,\beta}(b)^T J \, v(b, \lambda) - \psi_{b,\beta}(0)^T J \, v(0, \lambda) = (l - \lambda) \int_0^b \psi_{b,\beta}(x)^T W(x) \, v(x, \lambda) \, dx.$$

If now λ is an eigenvalue for the boundary-value problem (1.5.1), (4.3.2), then $v(\cdot, \lambda)$ will satisfy the same boundary condition at b as $\psi_{b,\beta}$ and hence

$$\psi_{b,\beta}(b)^T J \, v(b, \lambda) = 0.$$

Moreover, it is straightforward from (4.3.18) and (4.3.3) that

$$\psi_{b,\beta}(0)^T J \, v(0, \lambda) = 1.$$

Hence, for eigenvalues λ, the generalised Fourier transform (4.2.12) of $\psi_{b,\beta}$ is

$$\int_0^b \psi_{b,\beta}(t)^T W(t) v(t, \lambda) \, dt = \frac{1}{\lambda - l}.$$

Now (4.3.6) and the Parseval formula (4.2.13) give the important equality
$$\frac{\operatorname{Im} m_{b,\beta}(l)}{\operatorname{Im} l} = \int_0^b \psi_{b,\beta}^T W \overline{\psi_{b,\beta}} = \int_{\mathbb{R}} \frac{d\sigma_{b,\beta}(\lambda)}{|\lambda - l|^2}. \tag{4.3.19}$$

Now consider the choice $l = i$. By Lemma 4.3.2 (b), $|m_{b,\beta}(l)| \leq K$ for some constant K, any β and (say) $b > 1$. Hence. by (4.3.19),
$$0 < \int_{\mathbb{R}} \frac{d\sigma_{b,\beta}(\lambda)}{1 + \lambda^2} \leq K.$$

Also, since $\sigma_{b,\beta}(0) = 0$ by (4.2.14),
$$|\sigma_{b,\beta}(\Lambda)| \leq \operatorname{sgn} \Lambda \int_0^\Lambda \frac{1+\Lambda^2}{1+\lambda^2} d\sigma_{b,\beta}(\lambda) \leq (1+\Lambda^2) K \qquad (\Lambda \in \mathbb{R}), \tag{4.3.20}$$
independently of b (> 1) and β.

Hence we can apply Helly's Selection Theorem to the sequence $(\sigma_{b_n,\beta_n})_{n\in\mathbb{N}}$ with $b_n \to \infty$ ($n \to \infty$) and any β_n, obtaining the existence of a non-decreasing function $\sigma : \mathbb{R} \to \mathbb{R}$ which is the pointwise limit of a subsequence.

With σ so obtained, the next step of the proof is to establish (4.3.17) which expresses the m-function in terms of σ, and from which we will then show that the inverse relation (4.3.16) follows. An immediate consequence of (4.3.16) is that σ is independent of the choice of the sequences $(b_n)_{n\in\mathbb{N}}$ and $(\beta_n)_{n\in\mathbb{N}}$, which implies that $\lim_{b\to\infty} \sigma_{b,\beta} = \sigma$ pointwise, to complete the proof of part (a).

Let b_n, β_n be (sub-)sequences such that $\sigma_{b_n,\beta_n} \to \sigma$ ($n \to \infty$) pointwise. Then, by Helly's Integration Theorem,
$$\int_{-\Lambda}^\Lambda \frac{d\sigma(\lambda)}{1+\lambda^2} = \lim_{n\to\infty} \int_{-\Lambda}^\Lambda \frac{d\sigma_{b_n,\beta_n}(\lambda)}{1+\lambda^2} \leq K$$
for any $\Lambda > 0$, and so also
$$\int_{\mathbb{R}} \frac{d\sigma(\lambda)}{1+\lambda^2} \leq K.$$

Hence, for $\mu \geq 1$,
$$\left| \int_{|\lambda|\geq\mu} \frac{d\sigma_{b_n,\beta_n}(\lambda)}{\lambda^3} \right| \leq \int_{|\lambda|\geq\mu} \frac{2}{\mu(1+\lambda^2)} d\sigma_{b_n,\beta_n}(\lambda) \leq \frac{2K}{\mu}.$$

Therefore, for $\lambda', \lambda'' \in \mathbb{C} \setminus \mathbb{R}$, we have from (4.3.19)
$$\frac{\operatorname{Im} m(\lambda')}{\operatorname{Im} \lambda'} - \frac{\operatorname{Im} m(\lambda'')}{\operatorname{Im} \lambda''} = \lim_{n\to\infty} \int_{\mathbb{R}} \left(\frac{1}{|\nu - \lambda'|^2} - \frac{1}{|\nu - \lambda''|^2} \right) d\sigma_{b_n,\beta_n}(\nu)$$
$$= \lim_{n\to\infty} \int_{|\nu|\leq\mu} \left(\frac{1}{|\nu - \lambda'|^2} - \frac{1}{|\nu - \lambda''|^2} \right) d\sigma_{b_n,\beta_n}(\nu) + \frac{2CK}{\mu}$$
$$\to \int_{\mathbb{R}} \left(\frac{1}{|\nu - \lambda'|^2} - \frac{1}{|\nu - \lambda''|^2} \right) d\sigma(\nu)$$

4.3. The spectral function for the half-line problem

as $\mu \to \infty$, where C is a constant and we use Helly's Integration Theorem again. This gives (4.3.17), representing $\operatorname{Im} m$ as an integral with integrator σ.

For the converse (4.3.16), assume that $\lambda' < \lambda''$ are points where σ is continuous and let $\epsilon > 0$. Then

$$\int_{\lambda'}^{\lambda''} \operatorname{Im} m(t + i\epsilon)\, dt = \epsilon \int_{\lambda'}^{\lambda''} \int_{-\infty}^{\infty} \frac{d\sigma(\lambda)}{|\lambda - t - i\epsilon|^2}\, dt + \epsilon c\, (\lambda'' - \lambda')$$

$$= \int_{-\infty}^{\infty} \int_{\frac{\lambda' - \lambda}{\epsilon}}^{\frac{\lambda'' - \lambda}{\epsilon}} \frac{ds}{1 + s^2}\, d\sigma(\lambda) + \epsilon c\, (\lambda'' - \lambda'),$$

and observing that

$$\lim_{\epsilon \to 0} \int_{\frac{\lambda' - \lambda}{\epsilon}}^{\frac{\lambda'' - \lambda}{\epsilon}} \frac{ds}{1 + s^2} = \lim_{\epsilon \to 0} \left(\arctan \frac{\lambda'' - \lambda}{\epsilon} - \arctan \frac{\lambda' - \lambda}{\epsilon} \right) = \pi \chi_{(\lambda', \lambda'')}(\lambda)$$

uniformly outside any neighbourhood of $\{\lambda', \lambda''\}$ and that σ is continuous at λ', λ'', we obtain the *Stieltjes inversion formula* (4.3.16). □

The spectral function σ is the basis for formulating the Parseval and expansion formulae which correspond to (4.2.13) and (4.2.15) for the boundary-value problem comprising (1.5.1) on $[0, \infty)$ with the boundary condition (4.3.1) at 0. To develop these formulae, we first need some preliminary definitions and a lemma.

Returning to (4.2.3) with $\lambda = 0$, we define the operator $\mathcal{R}_0$ on $L^2([0, b]; W)$ such that $\mathcal{R}_0 h$ is the unique solution of the inhomogeneous initial-value problem

$$u' = JB\, u + JW\, h, \qquad u(0) = \begin{pmatrix} \sin \alpha \\ \cos \alpha \end{pmatrix}.$$

Moreover, we set $S_b := \{h \in L^2([0, b]; W) \mid \mathcal{R}_0 h(b) = 0\}$ and $D_b := \mathcal{R}_0 S_b$. Note that an element of D_{b_1} can be extended by 0 to give an element of D_{b_2} for any $b_2 > b_1$.

Lemma 4.3.6. D_b *is a dense subspace of* $L^2([0, b]; W)$.

Proof. We first show that the one-dimensional solution space

$$S'_b := \{g \in L^2([0, b]; W) \mid g' = JB\, g,\, g \text{ satisfies (4.3.1)}\}$$

is the orthogonal complement of S_b. Let $h \in L^2([0, b]; W)$ and $u := \mathcal{R}_0 h$. Then for any $g \in S'_b$,

$$(h, g) = \int_0^b h^T W \overline{g} = \int_0^b (-Ju' - Bu)^T \overline{g}$$
$$= u(b)^T J \overline{g(b)}$$

after an integration by parts. Thus h is orthogonal to S'_b if and only if $h \in S_b$. This means that $(S'_b)^\perp = S_b$ and, since S'_b is one-dimensional and therefore closed, $S_b^\perp = (S'_b)^{\perp\perp} = S'_b$.

Now let $F \in (\mathcal{R}_0\, S_b)^\perp$ and let U any solution of

$$U' = JBU + JWF$$

satisfying the boundary condition (4.3.1). Then, for any $h \in S_b$ and $u := \mathcal{R}_0\, h$, we have

$$(U, h) = \int_0^b U^T \overline{(-J u' - B u)} = \int_0^b F^T\, W\, \overline{u} = 0,$$

the integration by parts giving no boundary terms due to the properties of u. Since $h \in S_b$ was arbitrary, this shows that $U \in S_b^\perp = S'_b$. Hence $WF = 0$, which makes $F = 0$ in $L^2([0, b]; W)$. $\square$

Theorem 4.3.7. *Assume that the boundary-value problem* (1.5.1), (4.3.1) *on* $[0, \infty)$ *is in the limit-point case. Then for each* $f \in D_b \subset L^2([0, \infty); W)$, *with some* $b > 0$, *the generalised Fourier transform*

$$g(\lambda) = \int_0^\infty f(x)^T W(x) \overline{v(x, \lambda)} dx \quad (\lambda \in \mathbb{R})$$

satisfies $g \in L^2(\mathbb{R}; d\sigma)$ *and*

$$\int_0^\infty f^T\, W\, \overline{f} = \int_\mathbb{R} |g|^2\, d\sigma. \tag{4.3.21}$$

The mapping $f \mapsto g$ *extends to an isometry* $\mathcal{F}$ *between* $L^2([0, \infty); W)$ *and* $L^2(\mathbb{R}; d\sigma)$. *If* $f \in L^2([0, \infty); W)$ *and* $g := \mathcal{F} f$, *then*

$$f(x) = \lim_{\mu, \nu \to \infty} \int_{-\mu}^\nu g(\lambda)\, v(x, \lambda)\, d\sigma(\lambda), \tag{4.3.22}$$

where the limit is understood in the $L^2([0, \infty); W)$ *sense.*

Proof. Let $b > 0$ and $f \in D_b$. Then there exists $h \in S_b$ such that $f = \mathcal{R}_0\, h$, and so $f' = JBf + JWh$. Both f and h can be extended by 0 to the interval $[0, \infty)$, and then their restrictions to $[0, b']$ are in $D_{b'}$ and $S_{b'}$, respectively, for any $b' > b$.

The generalised Fourier transform of h is, for $\lambda \in \mathbb{R}$,

$$\int_0^\infty h^T\, W\, v(\cdot, \lambda) = \int_0^b (-J f' - B f)^T\, v(\cdot, \lambda)$$

$$= \lambda \int_0^b f^T\, W\, v(\cdot, \lambda) = \lambda g(\lambda), \tag{4.3.23}$$

4.3. The spectral function for the half-line problem

where g is the generalised Fourier transform of f. By Parseval's formula (4.2.13), we have

$$\int_{\mathbb{R}} \lambda^2 |g(\lambda)|^2 \, d\sigma_{b',\beta}(\lambda) = \int_0^\infty h^T W \overline{h} = C$$

for any $b' > b$ and any β, where C is independent of b' and β.

Using (4.2.13) again, it follows that

$$\left| \int_0^\infty f^T W \overline{f} - \int_{-\Lambda}^{\Lambda} |g(\lambda)|^2 \, d\sigma_{b',\beta}(\lambda) \right| \leq \int_{|\lambda|>\Lambda} \frac{\lambda^2}{\Lambda^2} |g(\lambda)|^2 \, d\sigma_{b',\beta}(\lambda) \leq \frac{C}{\Lambda^2}$$

for any $\Lambda > 0$. By Helly's Integration Theorem, we obtain

$$\left| \int_0^\infty f^T W \overline{f} - \int_{-\Lambda}^{\Lambda} |g(\lambda)|^2 \, d\sigma(\lambda) \right| \leq \frac{C}{\Lambda^2}$$

in the limit $b' \to \infty$, and hence (4.3.21) in the limit $\Lambda \to \infty$.

Since $\bigcup_{b>0} D_b$ is dense in $L^2([0,\infty); W)$ by Lemma 4.3.6, it follows immediately that the isometry extends to the whole Hilbert space.

For the proof of the expansion formula (4.3.22), let

$$f_{\mu,\nu}(x) := \int_{-\mu}^{\nu} g(\lambda) \, v(x,\lambda) \, d\sigma(\lambda) \qquad (x \in [0,\infty)).$$

For $b > 0$, let $h_b := (f - f_{\mu,\nu}) \chi_{[0,b]}$. Then, as $\mathcal{F}$ is an isometry,

$$(f, h_b) = \int_{\mathbb{R}} g(\lambda) \overline{\mathcal{F} h_b(\lambda)} \, d\sigma(\lambda);$$

also

$$(f_{\mu,\nu}, h_b) = \int_0^\infty \left(\int_{-\mu}^{\nu} g(\lambda) \, v(t,\lambda) \, d\sigma(\lambda) \right)^T W(t) \, \overline{h_b(t)} \, dt$$

$$= \int_{-\mu}^{\nu} g(\lambda) \int_0^\infty v(t,\lambda)^T W(t) \, \overline{h_b(t)} \, dt \, d\sigma(\lambda),$$

and so

$$(f - f_{\mu,\nu}, h_b) = \int_{\mathbb{R} \setminus (-\mu,\nu)} g(\lambda) \, \overline{\mathcal{F} h_b(\lambda)} \, d\sigma(\lambda).$$

Therefore, using the Cauchy-Schwarz inequality and the isometry property of $\mathcal{F}$,

$$\int_0^b (f - f_{\mu,\nu})^T W \overline{f - f_{\mu,\nu}} = (f - f_{\mu,\nu}, h_b)$$

$$\leq \left(\int_{\mathbb{R} \setminus (-\mu,\nu)} |g|^2 \, d\sigma \right)^{1/2} \left(\int_0^b (f - f_{\mu,\nu})^T W \overline{f - f_{\mu,\nu}} \right)^{1/2}.$$

Upon letting $b \to \infty$ and then $\mu, \nu \to \infty$, we obtain (4.3.22). $\square$

The generalised Fourier transform of a general element $f \in L^2([0,\infty); W)$ can be shown to take the form

$$(\mathcal{F}f)(\lambda) = \lim_{b\to\infty} \int_0^b f(x)^T \, W(x) \, \overline{v(x,\lambda)} \, dx \qquad (\lambda \in \mathbb{R}),$$

where the limit is understood in the $L^2(\mathbb{R}; d\sigma)$ sense.

4.4 Self-adjoint half-line operators

We now show how the generalised Fourier transform of Theorem 4.3.7 can be used to construct a self-adjoint linear operator associated with the boundary-value problem (1.5.1),(4.3.1). Moreover, we shall establish a link between the spectral properties of this operator and the growth of the spectral function, which will also afford a practical technique for the study of the spectrum.

The calculation in (4.3.23) shows that

$$(\mathcal{F}h)(\lambda) = \lambda\,(\mathcal{F}f)(\lambda)$$

for $f \in \bigcup_{b>0} D_b$ and $h \in L^2([0,\infty); W)$ such that $f' = JBf + JWh$. Thus the generalised Fourier transform relates the first-order system with boundary condition (4.3.1) to the operator of multiplication by the variable λ in the Hilbert space $L^2(\mathbb{R}; d\sigma)$. The natural domain for this multiplication operator $\mathcal{M}$ is the subspace of functions $y \in L^2(\mathbb{R}; d\sigma)$ such that

$$\int_\mathbb{R} |y(\lambda)|^2 (\lambda^2 + 1) \, d\sigma(\lambda) < \infty,$$

or equivalently the Hilbert space $L^2(\mathbb{R}; (\lambda^2 + 1) \, d\sigma(\lambda))$.

The generalised Fourier transforms of functions in $\bigcup_{b>0} D_b$ are dense in the latter Hilbert space; indeed if $y \in L^2(\mathbb{R}; (\lambda^2+1) \, d\sigma(\lambda))$ is orthogonal to $\mathcal{F}(\bigcup_{b>0} D_b)$, then

$$0 = \int_\mathbb{R} y(\lambda) \int_0^\infty \overline{u(x)}^T W(x) \, v(x,\lambda) \, dx \, (\lambda^2+1) \, d\sigma(\lambda)$$
$$= \int_0^\infty \overline{u(x)}^T W(x) \int_\mathbb{R} y(\lambda) v(x,\lambda) \, (\lambda^2+1) \, d\sigma(\lambda) \, dx$$

for all $u \in \bigcup_{b>0} D_b$. As the latter space is dense in $L^2([0,\infty); W)$ by Lemma 4.3.6, it follows that

$$\int_\mathbb{R} y(\lambda) v(\cdot, \lambda) \, (\lambda^2+1) \, d\sigma(\lambda) = 0 \qquad (4.4.1)$$

in $L^2([0,\infty); W)$, and so by Theorem 4.3.7 $y(\lambda)\,(\lambda^2+1) = 0$ in $L^2(\mathbb{R}; d\sigma)$. This in turn implies that $y(\lambda) = 0$ a.e. with respect to the measure $(\lambda^2+1) d\sigma(\lambda)$.

4.4. Self-adjoint half-line operators

Thus there is a closed operator
$$H_\alpha := \mathcal{F}^{-1} \mathcal{M} \mathcal{F}$$
in the Hilbert space $L^2([0,\infty); W)$, associated with the boundary-value problem (1.5.1), (4.3.1) on $[0,\infty)$; its domain is
$$D(H_\alpha) = \mathcal{F}^{-1} L^2(\mathbb{R}; (\lambda^2 + 1)\, d\sigma(\lambda)).$$

Although we define the operator H_α in terms of the generalised Fourier transform, for practical calculations it is often more convenient to know that it acts as a differential operator on a space of functions with easily verified properties. In order to characterise the differential operator and its domain, we begin with the observation that the elements of the domain of H_α arise as limits of regular functions of compact support. In fact, the following statement shows that the functions in D_b, with $b > 0$, form an *operator core* for H_α.

Lemma 4.4.1. *Let $u \in D(H_\alpha)$. Then there is a sequence $(u_n)_{n \in \mathbb{N}}$ in $\bigcup_{b>0} D_b$ such that both*
$$\lim_{n \to \infty} \|u - u_n\| = 0 \quad \text{and} \quad \lim_{n \to \infty} \|H_\alpha u - H_\alpha u_n\| = 0.$$

Proof. By definition, $u = \mathcal{F}^{-1} v$ with some $v \in L^2(\mathbb{R}; (\lambda^2 + 1) d\sigma(\lambda))$. By Lemma 4.3.6 and (4.3.21), there is a sequence $(v_n) \in \mathcal{F} \bigcup_{b>0} D_b$ such that
$$\lim_{n \to \infty} \int_\mathbb{R} |v(\lambda) - v_n(\lambda)|^2 (\lambda^2 + 1)\, d\sigma(\lambda) = 0.$$

This implies that both
$$\|u - u_n\|^2 = \int_\mathbb{R} |v(\lambda) - v_n(\lambda)|^2\, d\sigma(\lambda) \to \infty$$
and
$$\|H_\alpha u - H_\alpha u_n\|^2 = \int_\mathbb{R} |\lambda v(\lambda) - \lambda v_n(\lambda)|^2\, d\sigma(\lambda) \to \infty$$
as $n \to \infty$. $\square$

Before we proceed to investigate further properties of the operator H_α in the general form, a formal calculation can illustrate how it realises the Sturm-Liouville and Dirac operators specifically.

For the Sturm-Liouville case, $L^2([0,\infty); W)$ is identified with $L^2([0,\infty); w)$, the functions in the latter space forming the first components of elements of the former space. The generalised Fourier transform of $f \in L^2([0,\infty); w)$ for real λ becomes
$$(\mathcal{F} f)(\lambda) = \int_0^\infty f(x)\, w(x)\, v_1(x, \lambda)\, dx$$

and, since $\lambda w(x) v_1(x,\lambda) = -(p v_1')'(x,\lambda) + q(x) v_1(x,\lambda)$, two integrations by parts (without boundary terms) give

$$\lambda (\mathcal{F} f)(\lambda) = \int_0^\infty \{-(p f')'(x) + q(x) f(x))\} v_1(x,\lambda) \, dx,$$

which shows that

$$H_\alpha f = \mathcal{F}^{-1} M \mathcal{F} f = \frac{1}{w} ((-p f')' + q f). \tag{4.4.2}$$

The same calculation can be expressed in terms of the Sturm-Liouville system, complementing f with a second component $p f'$, with a single integration by parts as follows.

$$\lambda (\mathcal{F} f)(\lambda) = \int_0^\infty \begin{pmatrix} f \\ p f' \end{pmatrix}^T \lambda W v(\cdot, \lambda)$$

$$= \int_0^\infty \begin{pmatrix} f \\ p f' \end{pmatrix}^T \lambda (-J v'(\cdot,\lambda) - B v(\cdot,\lambda))$$

$$= \int_0^\infty \left(-J \begin{pmatrix} f \\ p f' \end{pmatrix}' - B \begin{pmatrix} f \\ p f' \end{pmatrix} \right)^T v(\cdot,\lambda).$$

In view of the definition of J and B, the second component of the bracket in the integrand vanishes, while the first component gives the generalised Fourier transform of (4.4.2).

In this case the elements of the operator core of Lemma 4.4.1 can be characterised as functions in

$$D_b = \{y \in L^2([0,\infty); w) \mid y, py' \in AC_{\text{loc}}([0,\infty)), y(x) = 0 \; (x \geq b),$$
$$y(0) \cos \alpha = p y'(0) \sin \alpha\},$$

where $b > 0$. Hence it follows that the operator domain is

$$D(H_\alpha) = \{y \in L^2([0,\infty); w) \mid y, py' \in AC_{\text{loc}}([0,\infty)), y(0) \cos \alpha = p y'(0) \sin \alpha,$$
$$\frac{1}{w}(-(pu')' + qu) \in L^2([0,\infty); w)\}. \tag{4.4.3}$$

For the Dirac system, we have $L^2([0,\infty); W) = L^2([0,\infty))^2$ and simply

$$\lambda \mathcal{F} f(\lambda) = \int_0^\infty f^T \lambda v(\cdot,\lambda) = \int_0^\infty f^T \left(-i\sigma_2 v'(\cdot,\lambda) + (p_1 \sigma_3 + p_2 \sigma_1 + q) v(\cdot,\lambda)\right)$$

$$= \int_0^\infty (-i\sigma_2 f' + (p_1 \sigma_3 + p_2 \sigma_1 + q) f)^T v(\cdot,\lambda).$$

Hence

$$H_\alpha f = \mathcal{F}^{-1} M \mathcal{F} f = -i\sigma_2 f' + (p_1 \sigma_3 + p_2 \sigma_1 + q) f.$$

4.4. Self-adjoint half-line operators

Here we have

$$D_b = \{u \in L^2([0,\infty); W) \mid u \in AC_{\text{loc}}([0,\infty)), u(x) = 0 \ (x \geq b),$$
$$u_1(0) \cos\alpha = u_2(0) \sin\alpha\}$$

and

$$D(H_\alpha) = \{u \in L^2([0,\infty); W) \mid u \in AC_{\text{loc}}([0,\infty)), u_1(0)\cos\alpha = u_2(0)\sin\alpha,$$
$$-i\sigma_2 u' + (p_1\sigma_3 + p_2\sigma_2 + q)u \in L^2([0,\infty); W)\}. \quad (4.4.4)$$

We now proceed with further properties of the operator H_α in the general setting, with the aim of establishing a connexion between its spectral properties and the behaviour of the spectral function.

Theorem 4.4.2. *The operator H_α is self-adjoint in $L^2([0,\infty); W)$.*

Proof. The multiplication operator $\mathcal{M}$ in $L^2(\mathbb{R}; d\sigma)$ has adjoint domain

$$D(\mathcal{M}^*) = \{u \in L^2(\mathbb{R}; d\sigma) \mid \exists v \in L^2(\mathbb{R}; d\sigma) \forall y \in D(\mathcal{M}) : (u, \mathcal{M}y) = (v, y)\}.$$

It is clear that if $u \in D(\mathcal{M})$, then $v := \mathcal{M}u$ will have the desired property, and so $\mathcal{M} \subset \mathcal{M}^*$. Moreover, if $u \in D(\mathcal{M})$ and $v = \mathcal{M}^* u$, then setting $y_n(\lambda) := \lambda \chi_{[-n,n]} u(\lambda)$ ($\lambda \in \mathbb{R}$) we have $y_n \in D(\mathcal{M})$ and

$$\int_{-n}^{n} \lambda^2 |u(\lambda)|^2 \, d\sigma(\lambda) = (u, \mathcal{M}(\lambda y_n)) = (v, \lambda y_n) = \int_{-n}^{n} v(\lambda) \lambda \overline{u}(\lambda) \, d\sigma(\lambda).$$

By the Cauchy-Schwarz inequality, this implies that

$$\int_{-n}^{n} \lambda^2 |u(\lambda)|^2 \, d\sigma(\lambda) \leq \|v\|^2,$$

and hence, sending n to infinity, that $u \in D(\mathcal{M})$.

Thus $\mathcal{M}$ is a self-adjoint operator. The self-adjointness of H_α now follows, as $\mathcal{F}$ is a unitary operator. $\square$

The *spectrum* of H_α is defined as the set of points $\mu \in \mathbb{C}$ such that $H_\alpha - \mu$ does not have a bounded inverse. In particular, this is the case if there is a $u \in D(H_\alpha) \setminus \{0\}$ such that $H_\alpha u = \mu u$; in this situation μ is called an *eigenvalue* of H_α and u a corresponding *eigenfunction*.

The relationship between the spectrum of H_α and the spectral function σ can be summarised as follows.

Theorem 4.4.3. *The spectrum of H_α is a subset of $\mathbb{R}$. A number $\mu \in \mathbb{R}$ is in the spectrum of H_α if and only if*

$$\sigma(\mu + \epsilon) > \sigma(\mu - \epsilon) \quad (4.4.5)$$

for all $\epsilon > 0$. Moreover, μ is an eigenvalue of H_α if and only if σ is discontinuous at μ.

Proof. We present the proof in five parts (a)-(e).

(a) If $\mu \in \mathbb{C}\setminus\mathbb{R}$, then for $f \in D(H_\alpha)$ we have $h := (H_\alpha - \mu)f \in L^2([0,\infty); W)$ and $\mathcal{F}h = (\lambda - \mu)\mathcal{F}f$. Hence

$$f = \mathcal{F}^{-1}\left(\frac{1}{\lambda - \mu}\mathcal{F}h\right)$$
$$= \int_\mathbb{R} \frac{1}{\lambda - \mu}\left(\int_0^\infty h(t)^T W(t) v(t, \lambda)\, dt\right) v(\cdot, \lambda)\, d\sigma(\lambda). \quad (4.4.6)$$

As

$$\left|\frac{1}{\lambda - \mu}\right| \leq \frac{1}{|\operatorname{Im}\mu|}$$

and $\mathcal{F}$ is an isometry,

$$\|f\| = \|\mathcal{F}f\| = \left\|\frac{1}{\lambda - \mu}\mathcal{F}h\right\| \leq \frac{1}{|\operatorname{Im}\mu|}\|\mathcal{F}h\| = \|h\|.$$

Hence $H_\alpha - \mu$ has a bounded inverse.

(b) Let $\mu \in \mathbb{R}$ and assume that there is an $\epsilon > 0$ such that $\sigma(\mu - \epsilon) = \sigma(\mu + \epsilon)$. Then for f and h as in (a) we have (4.4.6). The integrator of the outer integral gives no weight to the interval $(\mu - \epsilon, \mu + \epsilon)$, and

$$\frac{1}{\lambda - \mu} \leq \frac{1}{\epsilon}$$

outside this interval. Therefore

$$\|f\| = \|\mathcal{F}f\| = \left\|\frac{1}{\lambda - \mu}\mathcal{F}h\right\| \leq \frac{1}{\epsilon}\|\mathcal{F}h\| = \frac{1}{\epsilon}\|h\|,$$

and so $H_\alpha - \mu$ has a bounded inverse of operator norm no greater than $1/\epsilon$.

(c) Assume (4.4.5) is satisfied, so that $\delta_n := \sigma(\mu + \frac{1}{n}) - \sigma(\mu - \frac{1}{n}) > 0$ for all sufficiently large $n \in \mathbb{N}$. Let $g_n(\lambda) := \chi_{(\mu - \frac{1}{n}, \mu + \frac{1}{n})}/\sqrt{\delta_n} \in L^2(\mathbb{R}; d\sigma)$ and let

$$f_n(x) := \int_\mathbb{R} g_n(\lambda) v(x, \lambda)\, d\sigma(\lambda).$$

Then $\|f_n\| = \|g_n\| = 1$ and

$$\|(H_\alpha - \mu)f_n\|^2 = \int_\mathbb{R} |\lambda - \mu|^2 |g_n(\lambda)|^2\, d\sigma(\lambda)$$
$$= \frac{1}{\delta_n}\int_{\mu - \frac{1}{n}}^{\mu + \frac{1}{n}} |\lambda - \mu|^2\, d\sigma(\lambda) \leq \frac{1}{n^2}. \quad (4.4.7)$$

This implies that μ is in the spectrum of H_α because, if there were a bounded inverse $(H_\alpha - \mu)^{-1}$, then

$$1 = \|f_n\| \leq \|(H_\alpha - \mu)^{-1}\|\,\|(H_\alpha - \mu)f_n\| \to 0 \quad (n \to \infty).$$

4.4. Self-adjoint half-line operators

(d) Assume that μ is an eigenvalue of H_α and that $u \in D(H_\alpha)$ is an eigenfunction with $\|u\| = 1$. Then $g := \mathcal{F}u$ satisfies $\|g\| = 1$ and $(\lambda - \mu)\, g(\lambda) = \mathcal{F}(H_\alpha - \mu)\, u = 0$. Therefore

$$0 = \int_\mathbb{R} |\lambda - \mu|^2 |g(\lambda)|^2 \, d\sigma(\lambda) \geq \epsilon^2 \int_{\mathbb{R} \setminus (\mu - \epsilon, \mu + \epsilon)} |g|^2 \, d\sigma \geq 0,$$

and hence

$$\int_{\mu - \epsilon}^{\mu + \epsilon} |g|^2 \, d\sigma = 1$$

for all $\epsilon > 0$. This shows that σ must be discontinuous at μ.

(e) Assume σ has a saltus $s > 0$ at μ. Then $g := \chi_{\{\mu\}} \in L^2(\mathbb{R}; d\sigma)$ and $\|g\|^2 = s > 0$. Furthermore, $u := \mathcal{F}^{-1} g \in D(H_\alpha)$ gives $\|u\| = \|g\| > 0$ and

$$(H_\alpha - \mu)\, u = \mathcal{F}^{-1}(\lambda - \mu)\, g = 0.$$

Therefore μ is an eigenvalue of H_α. $\square$

Since the set of discontinuity points of any non-decreasing function is countable, Theorem 4.4.3 implies that H_α has at most countably many eigenvalues; this corresponds to the fact that their eigenfunctions can be chosen orthonormal in the separable Hilbert space $L^2([0, \infty); W)$.

The proof of part (c) above shows that points of the spectrum of H_α can be characterised by the existence of a *singular sequence* as follows.

Corollary 4.4.4. (a) *A number $\lambda \in \mathbb{C}$ is in the spectrum of H_α if and only if there exists a sequence $(f_n)_{n \in \mathbb{N}}$ in $D(H_\alpha)$ such that $\|f_n\| = 1$ and*

$$\lim_{n \to \infty} \|(H_\alpha - \lambda)\, f_n\| = 0.$$

(b) *The sequence in (a) can be chosen in $\bigcup_{b>0} D_b$ without loss of generality.*

Proof. We only need to show (b). Let $\lambda \in \mathbb{C}$ be a point of the spectrum of H_α and $(f_n)_{n \in \mathbb{N}}$ a corresponding sequence in $D(H_\alpha)$. Then, for each $n \in \mathbb{N}$, there is a $\tilde{f}_n \in \bigcup_{b>0} D_b$ such that $\|f_n - \tilde{f}_n\| < 1/n$ and $\|H_\alpha f_n - H_\alpha \tilde{f}_n\| < 1/n$. Hence

$$\|(H_\alpha - \lambda)\, \tilde{f}_n\| \leq \|H_\alpha f_n - H_\alpha \tilde{f}_n\| + |\lambda|\, \|f_n - \tilde{f}_n\| + \|(H_\alpha - \lambda)\, f_n\| \to 0$$

as $n \to \infty$. As $\lim_{n \to \infty} \|\tilde{f}_n\| = 1$, the same holds true for the normalised sequence $(\tilde{f}_n / \|\tilde{f}_n\|)_{n \in \mathbb{N}}$. $\square$

By the Lebesgue Decomposition Theorem the spectral function σ can be written as a sum of non-decreasing functions,

$$\sigma = \sigma_{pp} + \sigma_{ac} + \sigma_{sc},$$

where σ_{pp} is constant except at its points of discontinuity, σ_{ac} is locally absolutely continuous and σ_{sc} is continuous with vanishing derivative a.e. with respect to Lebesgue measure.

This decomposition gives rise to different types of spectra, corresponding to the sets of points of increase of each of the component functions, in analogy to the statement of Theorem 4.4.3. These are the *pure point spectrum* σ_{pp}, the *absolutely continuous spectrum* σ_{ac} and the *singular continuous spectrum* σ_{sc}. The absolutely continuous spectrum and the singular continuous spectrum taken together are called the *continuous spectrum*. Denoting the spectrum of H_α by $\sigma(H_\alpha)$, we clearly have
$$\sigma(H_\alpha) \subset \sigma_{pp}(H_\alpha) \cup \sigma_{ac}(H_\alpha) \cup \sigma_{sc}(H_\alpha);$$
moreover, $\sigma_p(H_\alpha) \subset \sigma_{pp}(H_\alpha)$, where
$$\sigma_p(H_\alpha) := \{\lambda \in \mathbb{R} \mid \lambda \text{ is an eigenvalue of } H_\alpha\}$$
is called the *point spectrum*. The sets $\sigma(H_\alpha), \sigma_{pp}(H_\alpha), \sigma_{ac}(H_\alpha), \sigma_{sc}(H_\alpha)$ are closed subsets of the real line, but $\sigma_p(H_\alpha)$ need not be closed. Moreover, the sets $\sigma_{pp}(H_\alpha)$, $\sigma_{ac}(H_\alpha)$ and $\sigma_{sc}(H_\alpha)$ are not disjoint in general.

Part (d) of the proof of Theorem 4.3.11 shows that eigenfunctions of H_α have generalised Fourier transforms concentrated at the eigenvalue. This means that μ is an eigenvalue if and only if $v(\cdot, \mu) \in L^2([0, \infty); W)$.

4.5 The spectrum of the periodic boundary-value problem on the half-line

Now consider the boundary-value problem (1.5.1), (4.3.1) on $[0, \infty)$ under the assumption that B and W are a-periodic. As in section 4.4, we denote by H_α the associated operator in $L^2([0, \infty); W)$. Also, $\mathcal{I}$ denotes the set of instability points, as in section 1.6.

Theorem 4.5.1. (a) *If λ is an eigenvalue of H_α, then $\lambda \in \mathcal{I}$. Moreover, for any $n \in \mathbb{N}$, λ is an eigenvalue of the boundary-value problem (1.5.1) on $[0, na]$ with boundary conditions*
$$u_1(0)\cos\alpha - u_2(0)\sin\alpha = 0, \qquad u_1(na)\cos\alpha - u_2(na)\sin\alpha = 0. \quad (4.5.1)$$

(b) *Each instability interval contains at most one eigenvalue and no other points of the spectrum of H_α.*

Proof. (a) Let $\lambda \in \mathbb{R}$ be an eigenvalue of H_α. Then the solution v defined in (4.2.10) satisfies $v(\cdot, \lambda) \in L^2([0, \infty); W)$. This is impossible if λ is a point of stability or conditional stability, since then every solution is a linear combination of Floquet solutions of periodic modulus, or of such a solution and a solution of linear growth, cf. (1.4.8) and (1.4.9).

4.5. The spectrum of the periodic boundary-value problem on the half-line

Hence λ must be a point of instability. In this case all solutions are linear combinations of the Floquet solutions $u_\pm$ given in (1.4.7); the square-integrability shows that
$$v(x,\lambda) = c e^{-\mu x/a} g_-(x) \qquad (x \geq 0),$$
with some constant $c \in \mathbb{C}$, $\operatorname{Re}\mu > 0$ and a-periodic g_-. In particular, $v(\cdot,\lambda)$ will satisfy the boundary conditions (4.5.1).

This makes λ an eigenvalue of the boundary-value problem with separated boundary conditions on $[0, na]$, with equal fixed boundary condition at both endpoints.

(b) By Theorem 2.4.5 (a), there is not more than one eigenvalue of the boundary-value problem (1.5.1), (4.5.1) in each instability interval; note that the a-periodic system is also na-periodic and, as instability is a global property of the set of solutions, the set $\mathcal{I}$ remains the same when na is taken as the period of the system.

Now let $[\lambda', \lambda''] \subset \mathcal{I}$ be an interval which does not contain an eigenvalue of H_α. Then by (a), the boundary-value problem (1.5.1), (4.5.1) has no eigenvalue in $[\lambda', \lambda'']$. Hence, by (4.2.14),
$$\sigma_{na,\alpha}(\lambda'') - \sigma_{na,\alpha}(\lambda') = 0$$
for all $n \in \mathbb{N}$. Passing to the limit $n \to \infty$, we obtain by Theorem 4.3.5 (a) that
$$\sigma(\lambda'') - \sigma(\lambda') = 0,$$
and so, by Theorem 4.4.3, H_α has no spectrum in $[\lambda', \lambda'']$. □

Now we turn to the stability set S of the periodic system (1.5.1) and our aim is to show that the spectral function σ is purely absolutely continuous on S. The following two lemmas prepare the way for the main result in Theorem 4.5.4.

Lemma 4.5.2. *Let S be the stability set of the periodic system (1.5.1) and let $[\lambda', \lambda''] \subset S$. Let $\rho(\cdot,\lambda)$ be the Prüfer radius, as defined in (2.2.1), of the solution $v(\cdot,\lambda)$ satisfying the initial condition (4.2.10) with parameter α. Then there is a constant $C \geq 1$ such that, for any $\alpha \in \mathbb{R}$,*
$$\frac{1}{C} \leq \rho(x,\lambda) \leq C \qquad (x \geq 0, \lambda \in [\lambda', \lambda'']).$$

Proof. Consider the monodromy matrix
$$M(\lambda) = \begin{pmatrix} \phi_{11}(\lambda) & \phi_{12}(\lambda) \\ \phi_{21}(\lambda) & \phi_{22}(\lambda) \end{pmatrix}$$
for $\lambda \in [\lambda', \lambda'']$. Then $\phi_{21}(\lambda) \neq 0$, since otherwise λ would be an eigenvalue of the boundary-value problem on $[0, a]$ with boundary conditions (2.3.2) with $\alpha = \beta = \pi/2$, and hence $\lambda \in \mathcal{I}$ by Theorem 2.4.5, contradicting the assumption about $[\lambda', \lambda'']$. By continuity, ϕ_{21} is bounded below by a positive constant on $[\lambda', \lambda'']$.

The eigenvalues of $M(\lambda)$ are $e^{\pm i\nu(\lambda)}$ with $\nu(\lambda) \in \mathbb{R}$, and its eigenvectors are

$$\begin{pmatrix} e^{i\nu(\lambda)} - \phi_{22}(\lambda) \\ \phi_{21}(\lambda) \end{pmatrix}, \quad \begin{pmatrix} e^{-i\nu(\lambda)} - \phi_{22}(\lambda) \\ \phi_{21}(\lambda) \end{pmatrix}. \tag{4.5.2}$$

We can write the initial vector

$$\begin{pmatrix} \sin\alpha \\ \cos\alpha \end{pmatrix} = \begin{pmatrix} e^{i\nu(\lambda)} - \phi_{22}(\lambda) & e^{-i\nu(\lambda)} - \phi_{22}(\lambda) \\ \phi_{21}(\lambda) & \phi_{21}(\lambda) \end{pmatrix} \xi(\lambda),$$

with coefficient vector

$$\xi(\lambda) = \frac{1}{(e^{i\nu(\lambda)} - e^{-i\nu(\lambda)})\phi_{21}(\lambda)} \begin{pmatrix} \phi_{21}(\lambda) & -(e^{-i\nu(\lambda)} - \phi_{22}(\lambda)) \\ -\phi_{21}(\lambda) & e^{i\nu(\lambda)} - \phi_{22}(\lambda) \end{pmatrix} \begin{pmatrix} \sin\alpha \\ \cos\alpha \end{pmatrix}.$$

As the difference of the Floquet multipliers $|e^{i\nu} - e^{-i\nu}|$ is bounded below by a positive constant, $\xi(\lambda)$ is bounded uniformly with respect to $\lambda \in [\lambda', \lambda'']$ and α. Hence

$$v(nL, \lambda) = M(\lambda)^n v(0, \lambda)$$
$$= e^{i\nu n} \begin{pmatrix} e^{i\nu(\lambda)} - \phi_{22}(\lambda) \\ \phi_{21}(\lambda) \end{pmatrix} \xi_1(\lambda) + e^{-i\nu n} \begin{pmatrix} e^{-i\nu(\lambda)} - \phi_{22}(\lambda) \\ \phi_{21}(\lambda) \end{pmatrix} \xi_2(\lambda),$$

which implies that

$$\rho(nL, \lambda) \leq \left| \begin{pmatrix} e^{i\nu(\lambda)} - \phi_{22}(\lambda) \\ \phi_{21}(\lambda) \end{pmatrix} \right| |\xi_1(\lambda)| + \left| \begin{pmatrix} e^{-i\nu(\lambda)} - \phi_{22}(\lambda) \\ \phi_{21}(\lambda) \end{pmatrix} \right| |\xi_2(\lambda)|.$$

Consequently, $\rho(nL, \lambda)$ is bounded uniformly with respect to $n \in \mathbb{N}$, $\alpha \in \mathbb{R}$ and $\lambda \in [\lambda', \lambda'']$.

For $x \in (nL, (n+1)L)$, (2.2.3) gives

$$\rho(x, \lambda) = \rho(nL, \lambda) \exp\left(\int_{nL}^{x} \begin{pmatrix} -\cos\theta \\ \sin\theta \end{pmatrix}^T (B + \lambda W) \begin{pmatrix} \sin\theta \\ \cos\theta \end{pmatrix} \right)$$

and, as the integral is uniformly bounded, we obtain the existence of a constant $C > 0$ such that

$$\rho(x, \lambda) < C \quad (x \geq 0, \lambda \in [\lambda', \lambda'']).$$

For the lower bound, we observe that the solution $u(\cdot, \lambda)$ with initial condition as in (4.2.10) has Prüfer radius $\tilde{\rho}(\cdot, \lambda)$ bounded by C, because it turns into the above $v(\cdot, \lambda)$ when α is replaced with $\alpha + \pi/2$. The constant Wronskian of u and v has square

$$1 = \begin{vmatrix} u_1 & v_1 \\ u_2 & v_2 \end{vmatrix}^2 = u_1^2 v_2^2 - 2u_1 u_2 v_1 v_2 + u_2^2 v_1^2,$$

4.5. The spectrum of the periodic boundary-value problem on the half-line

and so
$$C^2 \rho^2 \geq \tilde{\rho}^2 \rho^2 = u_1^2 v_1^2 + u_2^2 v_1^2 + u_1^2 v_2^2 + u_2^2 v_2^2$$
$$= 1 + (u_1 v_1 + u_2 v_2)^2 \geq 1.$$

□

For the Sturm-Liouville system, we need the following additional observation.

Lemma 4.5.3. *Let S be the stability set of the periodic Sturm-Liouville system and let $[\lambda', \lambda''] \subset S$. Then there is a constant $C' > 0$ such that, for any $\alpha \in [0, \pi)$, the solution $v(\cdot, \lambda)$ with initial condition (4.2.10) with parameter α satisfies*
$$\int_{(n-1)a}^{na} |v_1(x, \lambda)|^2 w(x)\, dx \geq C'$$
for all $n \in \mathbb{N}$ and $\lambda \in [\lambda', \lambda'']$.

Proof. Let $u(\cdot, \lambda)$ and $\overline{u(\cdot, \lambda)}$ be the Floquet solutions arising from the eigenvectors of the monodromy matrix (4.5.2). As they are linearly independent,
$$v(\cdot, \lambda) = c(\lambda)\, u(\cdot, \lambda) + d(\lambda)\, \overline{u(\cdot, \lambda)},$$
and as $v(\cdot, \lambda)$ is real-valued, $d = \bar{c}$. Also, $|c| > 0$ uniformly with respect to λ and α. Using (1.3.4), we find that
$$\int_{(n-1)a}^{na} |v_1(x, \lambda)|^2 w(x)\, dx = \int_{(n-1)a}^{na} v(x, \lambda)^T W(x) \overline{v(x, \lambda)}\, dx$$
$$= 2|c(\lambda)|^2 \int_0^a u(x, \lambda)^T W(x) \overline{u(x, \lambda)}\, dx$$
$$+ 2\operatorname{Re}\left(c(\lambda)^2 e^{2\nu(\lambda)n} \int_0^a u(x, \lambda)^T W(x) u(x, \lambda)\, dx\right). \tag{4.5.3}$$

Now
$$\left(\int_0^a u^T W \bar{u}\right)^2 - \left|\int_0^a u^T W u\right|^2 = \left(\int_0^a |u_1|^2 w\right)^2 - \left|\int_0^a u_1^2 w\right|^2$$
$$= 4\left(\int_0^a (\operatorname{Re} u_1)^2 w \int_0^a (\operatorname{Im} u_1)^2 w - \left(\int_0^a (\operatorname{Re} u_1)(\operatorname{Im} u_1)\, w\right)^2\right) \geq 0$$

by the Cauchy-Schwarz inequality, and equality would hold only if $\operatorname{Re} u_1$ and $\operatorname{Im} u_1$ were linearly dependent, which would contradict the linear independence of u and $\bar{u}$. Hence
$$\int_0^a u(x, \lambda)^T W(x) \overline{u(x, \lambda)}\, dx > \left|\int_0^a u(x, \lambda)^T W u(x, \lambda)\, dx\right|$$
for all $\lambda \in [\lambda', \lambda'']$, and by continuity the difference is bounded below by a positive constant. Thus (4.5.3) gives the required result. □

Theorem 4.5.4. *The spectral function for the boundary-value problem on $[0, \infty)$ for the Sturm-Liouville or Dirac system with periodic coefficients is purely absolutely continuous and strictly increasing on $\mathcal{S}$. The closure of $\mathcal{S}$ is purely absolutely continuous spectrum of H_α.*

Proof. Let $[\lambda', \lambda''] \subset \mathcal{S}$. The spectral function of the boundary-value problem on $[0, na]$ with boundary conditions (4.3.2) at 0 and at $b = na$ satisfies

$$\sigma_{na,\beta}(\lambda'') - \sigma_{na,\beta}(\lambda') = \sum_\lambda \frac{1}{\int_0^{na} v(x,\lambda)^T W(x) v(x,\lambda) \, dx}, \tag{4.5.4}$$

where the sum is taken over all eigenvalues $\lambda \in (\lambda', \lambda'']$ of this boundary-value problem, and $v(\cdot, \lambda)$ is the solution with initial value (4.2.10).

Consider a fixed n for the moment. Let $\lambda_J(\beta), \lambda_{J+1}(\beta), \ldots, \lambda_K(\beta)$ be the eigenvalues in $[\lambda', \lambda'']$ of the boundary-value problem with parameter β. Consider the Prüfer angle $\theta(\cdot, \lambda)$ of $v(\cdot, \lambda)$, with $\theta(0, \lambda) = \alpha$. Then, by Corollary 2.3.3, $\theta(na, \lambda)$ is monotone increasing in λ and

$$\theta(na, \lambda_r(\beta)) = \beta + r\pi.$$

Consequently, $\lambda_r(\beta)$ increases from $\lambda_{r-1}(\pi)$ to $\lambda_r(\pi)$ as β runs through $[0, \pi]$. Integrating (4.5.4) with respect to β and changing variables, we obtain

$$\int_0^\pi (\sigma_{na,\beta}(\lambda'') - \sigma_{na,\beta}(\lambda')) \, d\beta$$

$$= \int_{\lambda_J^{-1}(\lambda')}^{\pi} \frac{1}{\int_0^{na} v(\cdot, \lambda_J(\beta))^T W v(\cdot, \lambda_J(\beta))} \, d\beta$$

$$+ \int_0^\pi \frac{1}{\int_0^{na} v(\cdot, \lambda_{J+1}(\beta))^T W v(\cdot, \lambda_{J+1}(\beta))} \, d\beta$$

$$+ \cdots + \int_0^{\lambda_K^{-1}(\lambda'')} \frac{1}{\int_0^{na} v(\cdot, \lambda_K(\beta))^T W v(\cdot, \lambda_K(\beta))} \, d\beta$$

$$= \int_{\lambda'}^{\lambda''} \frac{1}{\int_0^{na} v(\cdot, \lambda)^T W v(\cdot, \lambda)} \frac{\partial \theta(na, \lambda)}{\partial \lambda} \, d\lambda. \tag{4.5.5}$$

Now, differentiating (2.2.2) with respect to λ and using (2.2.3), we find that $\dfrac{\partial \theta(\cdot, \lambda)}{\partial \lambda}$ satisfies the linear first-order differential equation

$$\left(\frac{\partial \theta(\cdot, \lambda)}{\partial \lambda}\right)' + 2(\log \rho(\cdot, \lambda))' \left(\frac{\partial \theta(\cdot, \lambda)}{\partial \lambda}\right) = \begin{pmatrix} \sin \theta(\cdot, \lambda) \\ \cos \theta(\cdot, \lambda) \end{pmatrix}^T W \begin{pmatrix} \sin \theta(\cdot, \lambda) \\ \cos \theta(\cdot, \lambda) \end{pmatrix},$$

where $\rho(\cdot, \lambda)$ is the Prüfer radius of $v(\cdot, \lambda)$. As $\dfrac{\partial \theta(0, \lambda)}{\partial \lambda} = 0$, it follows that

$$\left(\frac{\partial \theta(x, \lambda)}{\partial \lambda}\right) = \frac{1}{\rho(x, \lambda)^2} \int_0^x v(t, \lambda)^T W(t) v(t, \lambda) \, dt. \tag{4.5.6}$$

4.5. The spectrum of the periodic boundary-value problem on the half-line

Therefore (4.5.5) gives

$$\int_0^\pi (\sigma_{na,\beta}(\lambda'') - \sigma_{na,\beta}(\lambda'))\, d\beta = \int_{\lambda'}^{\lambda''} \frac{d\lambda}{\rho(na,\lambda)^2}.$$

By Lemma 4.2.2, this shows that

$$\frac{\lambda'' - \lambda'}{C^2} \leq \int_0^\pi (\sigma_{na,\beta}(\lambda'') - \sigma_{na,\beta}(\lambda'))\, d\beta \leq C^2(\lambda'' - \lambda'). \tag{4.5.7}$$

We now aim to let $n \to \infty$ in (4.5.7) and, to justify this process, we need the integrand in (4.5.7) to be uniformly bounded with respect to $n \in \mathbb{N}$ and $\beta \in [0, \pi]$. To see this, we use (4.5.6) to obtain

$$(K - J)\pi = \theta(na, \lambda_K(\beta)) - \theta(na, \lambda_J(\beta))$$
$$= \int_{\lambda_J(\beta)}^{\lambda_K(\beta)} \frac{\int_0^{na} v(\cdot, \lambda)^T\, W\, v(\cdot, \lambda)}{\rho(na, \lambda)^2}\, d\lambda$$
$$\leq (\lambda_K(\beta) - \lambda_J(\beta))\, naC^4 \leq (\lambda'' - \lambda')\, naC^4.$$

Then, using the fact that

$$\int_0^{na} v(\cdot, \lambda)^T\, W\, v(\cdot, \lambda) \geq nC'$$

(which follows from Lemma 4.5.3 for the Sturm-Liouville equation and from Lemma 4.5.2, with $C' = a/C^2$, for the Dirac system), we find that

$$\sigma_{na,\beta}(\lambda'') - \sigma_{na,\beta}(\lambda') \leq \frac{K - J + 1}{n/C'} \leq (\lambda'' - \lambda')\, \frac{aC^4 C'}{\pi} + \frac{C'}{n}.$$

From Theorem 4.3.10 (a) we know that $\sigma_{na,\beta} \to \sigma$ as $n \to \infty$, and Lebesgue's dominated convergence theorem gives

$$\lim_{n \to \infty} \int_0^\pi (\sigma_{na,\beta}(\lambda'') - \sigma_{na,\beta}(\lambda'))\, d\beta = \pi\, (\sigma(\lambda'') - \sigma(\lambda')).$$

When combined with (4.5.7), this gives

$$\frac{\lambda'' - \lambda'}{\pi C^2} \leq \sigma(\lambda'') - \sigma(\lambda') \leq \frac{C^2}{\pi}(\lambda'' - \lambda'). \tag{4.5.8}$$

Thus the spectral function is locally Lipschitz continuous, and hence absolutely continuous, as well as being strictly increasing in $\mathcal{S}$. $\square$

4.6 The spectral matrix for the full-line problem

The fundamental ideas for the construction of a spectral function in section 4.3 also apply when the equation (1.5.1) is considered on the whole real line. In this case, we first study the equation on the finite interval $[c, b]$ with separated boundary conditions

$$u_1(c)\cos\alpha - u_2(c)\sin\alpha = 0, \tag{4.6.1}$$
$$u_1(b)\cos\beta - u_2(b)\sin\beta = 0, \tag{4.6.2}$$

and then perform the double limiting process $b \to \infty$, $c \to -\infty$. Weyl's limit-point or limit-circle classification applies to the positive and the negative half-axes separately, and we shall focus on the situation where both $+\infty$ and $-\infty$ are in the limit-point case. This is the case for the a-periodic equation, by Proposition 4.3.4 for the end-point $+\infty$, and for the same reason at the end-point $-\infty$.

There is, however, the complication that, while the representation (4.3.22) was based on the particular solution $v(\cdot, \lambda)$ which satisfies the prescribed boundary condition at the finite left-hand end-point 0, there is no such distinguished solution for the whole-line problem. Consequently, the expansion formula must use a full fundamental system of (1.5.1) as a basis and, corresponding to the two solutions in the fundamental system, the generalised Fourier transform will be a two-component function.

Specifically, let $\Phi(\cdot, \lambda)$ be the canonical fundamental matrix of (1.5.1). Then, keeping $c < 0$ and $b > 0$ fixed for the moment, the orthonormal eigenfunctions of the boundary-value problem (1.5.1), (4.6.1), (4.6.2) can be written in the form $\Phi(\cdot, \lambda_j)\gamma_j$, where λ_j is the j-th eigenvalue, $j \in \mathcal{J}$, and $\gamma_j \in \mathbb{C}^2$. The Parseval formula (4.2.8) for a function $f \in L^2([c, b]; W)$ then takes the form

$$\int_c^b f^T W \overline{f} = \sum_{j \in \mathcal{J}} \left| \int_c^b f^T W \overline{\Phi(\cdot, \lambda_j)\gamma_j} \right|^2$$
$$= \int_\mathbb{R} g(\lambda)^T \, d\sigma_{c,b,\alpha,\beta}(\lambda)\, g(\lambda), \tag{4.6.3}$$

where we define the generalised Fourier transform

$$g(\lambda) := \left(\int_c^b f^T W \overline{\Phi(\cdot, \lambda)} \right)^T \qquad (\lambda \in \mathbb{R})$$

and the matrix-valued spectral function

$$\sigma_{c,b,\alpha,\beta}(\lambda) = \begin{cases} \sum\limits_{\lambda_j \in (0, \lambda]} \overline{\gamma_j}\, \gamma_j^T & \text{if } \lambda \geq 0, \\ -\sum\limits_{\lambda_j \in (\lambda, 0]} \overline{\gamma_j}\, \gamma_j^T & \text{if } \lambda < 0, \end{cases} \tag{4.6.4}$$

4.6. The spectral matrix for the full-line problem

in analogy to (4.2.14).

The integrator $\sigma_{c,b,\alpha,\beta}(\lambda)$ is written in the middle of the integral in (4.6.3) due to its matrix character. As each matrix $\overline{\gamma_j}\gamma_j^T$ is positive definite, $\sigma_{c,b,\alpha,\beta}(\lambda)$ is a monotone increasing 2×2 matrix-valued function in the sense that

$$\sigma_{c,b,\alpha,\beta}(\lambda') - \sigma_{c,b,\alpha,\beta}(\lambda)$$

is a positive semidefinite matrix whenever $\lambda < \lambda'$.

On the part interval $[0, b]$, the two columns of Φ coincide with the solutions u and v considered in section 4.3, with the boundary condition $v(0) = 0$. Thus, for non-real λ, there exists a unique $m_{b,\beta}(\lambda) \in \mathbb{C}$ such that the solution $\Phi(\cdot, \lambda)\begin{pmatrix} 1 \\ m_{b,\beta}(\lambda) \end{pmatrix}$ satisfies the boundary condition (4.6.2) at b; see (4.3.4). In like manner, there exists a unique $n_{c,\alpha}(\lambda) \in \mathbb{C}$ such that the solution $\Phi(\cdot, \lambda)\begin{pmatrix} 1 \\ n_{c,\alpha}(\lambda) \end{pmatrix}$ satisfies the boundary condition (4.6.1) at c.

We now wish to establish a connexion of the type of (4.3.19) between the spectral matrix $\sigma_{c,b,\alpha,\beta}$ and the two Titchmarsh-Weyl functions $m_{b,\beta}$ and $n_{c,\alpha}$. To this end, we fix $l \in \mathbb{C}$, $\operatorname{Im} l > 0$, and apply the Parseval formula (4.6.3) to the function

$$f(x) = \begin{cases} C_- \Phi(x, l)\begin{pmatrix} 1 \\ n \end{pmatrix} & \text{if } x < 0, \\ C_+ \Phi(x, l)\begin{pmatrix} 1 \\ m \end{pmatrix} & \text{if } x > 0, \end{cases}$$

where we abbreviate $n := n_{c,\alpha}(l)$, $m := m_{b,\beta}(l)$, and C_+, C_- are arbitrary constants, which will be chosen later. Then clearly $f \in L^2([c, d]; W)$ and

$$\int_c^b f^T W \overline{f} = \int_c^0 |C_-|^2 \begin{pmatrix} 1 \\ n \end{pmatrix}^T \Phi(x, l)^T W(x) \overline{\Phi(x, l)} \begin{pmatrix} 1 \\ \overline{n} \end{pmatrix}$$
$$+ \int_0^b |C_+|^2 \begin{pmatrix} 1 \\ m \end{pmatrix}^T \Phi(x, l)^T W(x) \overline{\Phi(x, l)} \begin{pmatrix} 1 \\ \overline{m} \end{pmatrix}$$
$$= \frac{|C_+|^2 \operatorname{Im} m - |C_-|^2 \operatorname{Im} n}{\operatorname{Im} l}, \qquad (4.6.5)$$

where we have used (4.3.6).

Since f satisfies both boundary conditions (4.6.1) and (4.6.2), and $u_j(0) = \gamma_j$, Lemma 4.3.1 gives

$$\int_c^b f^T W \overline{u_j} = \int_c^0 C_- \begin{pmatrix} 1 \\ n \end{pmatrix}^T \Phi(\cdot, l)^T W \overline{u_j}\, dx + \int_0^b C_+ \begin{pmatrix} 1 \\ m \end{pmatrix}^T \Phi(\cdot, l)^T W \overline{u_j}\, dx$$
$$= \frac{1}{l - \lambda_j} \begin{pmatrix} C_+ m - C_- n \\ C_- - C_+ \end{pmatrix}^T \overline{\gamma_j}.$$

Together with (4.6.5), this shows that the Parseval formula (4.6.3) yields

$$\frac{|C_+|^2 \operatorname{Im} m - |C_-|^2 \operatorname{Im} n}{\operatorname{Im} l} = \begin{pmatrix} C_+ m - C_- n \\ C_- - C_+ \end{pmatrix}^T \int_{\mathbb{R}} \frac{d\sigma_{c,b,\alpha,\beta}(\lambda)}{|l - \lambda|^2} \overline{\begin{pmatrix} C_+ m - C_- n \\ C_- - C_+ \end{pmatrix}} \tag{4.6.6}$$

for all $\begin{pmatrix} C_+ \\ C_- \end{pmatrix} \in \mathbb{C}^2$. This identity allows us to express the matrix

$$\Sigma(l) := \int_{\mathbb{R}} \frac{d\sigma_{c,b,\alpha,\beta}(\lambda)}{|l - \lambda|^2}$$

in terms of m and n as follows. The choice $C_- = C_+ = 1$ in (4.6.6) yields

$$\Sigma_{11} \operatorname{Im} l = \frac{\operatorname{Im}(m-n)}{|m-n|^2} = \operatorname{Im} \frac{1}{n-m}; \tag{4.6.7}$$

similarly, choosing $C_+ = n$, $C_- = m$, we find

$$\Sigma_{22} \operatorname{Im} l = \frac{\operatorname{Im}(m|n|^2 - n|m|^2)}{|m-n|^2} = \operatorname{Im} \frac{mn}{n-m}. \tag{4.6.8}$$

To determine the off-diagonal entries of Σ, we note that $\Sigma_{21} = \overline{\Sigma_{12}}$, since σ is symmetric. Choosing $C_+ = 1 + n$, $C_- = 1 + m$, we obtain

$$(\Sigma_{11} + \Sigma_{12} + \Sigma_{21} + \Sigma_{22}) \operatorname{Im} l$$
$$= \frac{(1 + 2\operatorname{Re} n + |n|^2) \operatorname{Im} m - (1 + 2\operatorname{Re} m + |m|^2) \operatorname{Im} n}{|m-n|^2}$$

and hence, using (4.6.7) and (4.6.8),

$$\operatorname{Re} \Sigma_{12} \operatorname{Im} l = \frac{\operatorname{Im} mn}{|m-n|^2}. \tag{4.6.9}$$

Similarly, the choice $C_+ = 1 + in$, $C_- = 1 + im$ gives

$$(\Sigma_{11} - i\Sigma_{12} + i\Sigma_{21} + \Sigma_{22}) \operatorname{Im} l$$
$$= \frac{(1 - 2\operatorname{Im} n + |n|^2) \operatorname{Im} m - (1 - 2\operatorname{Im} m + |m|^2) \operatorname{Im} n}{|m-n|^2},$$

and so $\operatorname{Im} \Sigma_{12} = 0$.

Collecting (4.6.7), (4.6.8) and (4.6.9) together, we thus obtain a representation analogous to (4.3.17),

$$\int_{-\infty}^{\infty} \frac{d\sigma_{c,b,\alpha,\beta}(\lambda)}{|\lambda - l|^2} = \frac{\operatorname{Im} M_{c,b,\alpha,\beta}(l)}{\operatorname{Im} l}$$

4.6. The spectral matrix for the full-line problem

with

$$M_{c,b,\alpha,\beta}(l) = \begin{pmatrix} \frac{m-n}{|m-n|^2} & \frac{mn}{|m-n|^2} \\ \frac{mn}{|m-n|^2} & \frac{m|n|^2 - n|m|^2}{|m-n|^2} \end{pmatrix}.$$

Now, keeping l fixed in the complex upper half-plane and changing the endpoints c and b, we know from Lemma 4.3.2 that m will lie in a fixed circle in the complex upper half-plane for all $b > 1$ and, by analogous considerations, that n will lie in a fixed circle in the complex lower half-plane for all $c < -1$. Hence $|m-n|$ will be bounded below by a positive constant while $|m|$ and $|n|$ are bounded above, uniformly in b and c. In particular, taking $l = i$ we find that there is a constant $K > 0$ such that

$$\int_{-\infty}^{\infty} \frac{d(\sigma_{c,b,\alpha,\beta})_{jj}(\lambda)}{1+\lambda^2} \leq K \quad (j \in \{1,2\}).$$

Note that $(\sigma_{c,b,\alpha,\beta})_{11}$ and $(\sigma_{c,b,\alpha,\beta})_{22}$ are real-valued, non-decreasing functions. The off-diagonal entries $(\sigma_{c,b,\alpha,\beta})_{12}$ and $(\sigma_{c,b,\alpha,\beta})_{21} = \overline{(\sigma_{c,b,\alpha,\beta})_{12}}$ do not necessarily have these properties, but they are still functions of locally bounded variation; indeed, the variation of σ_{12} over an interval (λ, λ') is

$$\operatorname*{Var}_{(\lambda,\lambda')} (\sigma_{c,b,\alpha,\beta})_{12} = \sum_j \left| \overline{(\gamma_j)_1} (\gamma_j)_2 \right| \leq \frac{1}{2} \sum_j \left(|(\gamma_j)_1|^2 + |(\gamma_j)_2|^2 \right)$$

$$= \frac{1}{2} \left(\operatorname*{Var}_{(\lambda,\lambda')} (\sigma_{c,b,\alpha,\beta})_{11} + \operatorname*{Var}_{(\lambda,\lambda')} (\sigma_{c,b,\alpha,\beta})_{22} \right),$$

where the sums are taken over all indices of eigenvalues in (λ, λ').

Thus the variation of the spectral matrix satisfies

$$\int_{-\infty}^{\infty} \frac{|d\sigma_{c,b,\alpha,\beta}(\lambda)|}{1+\lambda^2} \leq K.$$

Therefore Helly's Selection Theorem is applicable, ensuring convergence of the spectral matrices to a non-decreasing symmetric matrix-valued function σ along a subsequence as $b \to \infty$, $c \to -\infty$. As in the proof of Theorem 4.3.5, a Stieltjes inversion formula

$$\sigma(\lambda'') - \sigma(\lambda') = \lim_{\epsilon \to 0} \frac{1}{\pi} \int_{\lambda'}^{\lambda''} \operatorname{Im} M(t + i\epsilon) \, dt \qquad (4.6.10)$$

can be derived, where M is the limit of $M_{c,b,\alpha,\beta}$ as $\beta \to \infty$ and $\alpha \to -\infty$. This limit is unique because, by assumption, both ∞ and $-\infty$ are in the limit-point case. Hence (4.6.10) shows that the limiting spectral matrix σ is uniquely defined at its points of continuity.

The Parseval and expansion formulae for this spectral matrix can be derived as in the proof of Theorem 4.3.7. We only briefly sketch the modifications necessary in the present situation.

We now define the operator $\mathcal{R}_0$ on $L^2([c,b];W)$ such that $\mathcal{R}_0 h$ is the unique solution of the inhomogeneous initial-value problem

$$u' = JBu + JWh, \qquad u(c) = 0,$$

and set $S_{cb} := \{h \in L^2([c,b];W) \mid \mathcal{R}_0 h(b) = 0\}$ and $D_{cb} := \mathcal{R}_0 S_b$. By extension by 0, we can embed $D_{c_1 b_1}$ into $D_{c_2 b_2}$ if $(c_1, b_1) \subset (c_2, b_2)$.

Lemma 4.6.1. D_{cb} *is a dense subspace of* $L^2([c,b];W)$.

Proof. By the same calculation as in the proof of Lemma 4.3.6, the two-dimensional solution space

$$S'_{cb} := \{g \in L^2([c,b];W) \mid g' = JBg\}$$

is the orthogonal complement of S_{cb}. The density of D_{cb} then follows as before, replacing S_b with S_{cb}. □

We now define the generalised Fourier transform first for compactly supported functions $f \in D_{cb} \subset L^2(\mathbb{R};W)$ with arbitrary $c < b$ as

$$g(\lambda) := \left(\int_{-\infty}^{\infty} f^T W \overline{\Phi(\cdot, \lambda)} \right)^T \qquad (\lambda \in \mathbb{R}) \tag{4.6.11}$$

and then extend it to the whole Hilbert space. We write $L^2(\mathbb{R}; d\sigma)$ for the Hilbert space of two-component functions $g : \mathbb{R} \to \mathbb{C}^2$ with inner product

$$(g, h) := \int_{\mathbb{R}} g(\lambda)^T d\sigma(\lambda) \overline{h(\lambda)}.$$

Theorem 4.6.2. *Assume that* (1.5.1) *on* $\mathbb{R}$ *is in the limit-point case at both ends. Then, for each* $f \in D_{cb} \subset L^2([c,b];W)$, *with some* $c < b$, *the generalised Fourier transform* (4.6.11) *satisfies* $g \in L^2(\mathbb{R}; d\sigma)$ *and*

$$\int_{-\infty}^{\infty} f^T W \overline{f} = \int_{\mathbb{R}} g^T d\sigma \overline{g}.$$

The mapping $f \mapsto g$ *extends to an isometry* $\mathcal{F}$ *between* $L^2(\mathbb{R};W)$ *and* $L^2(\mathbb{R}; d\sigma)$. *If* $f \in L^2(\mathbb{R};W)$ *and* $g := \mathcal{F}f$, *then*

$$f(x) = \lim_{\mu,\nu \to \infty} \int_{-\mu}^{\nu} \Phi(x, \lambda) \, d\sigma(\lambda)^T g(\lambda), \tag{4.6.12}$$

where the limit is understood in the $L^2(\mathbb{R};W)$ *sense.*

Proof. The proof proceeds exactly as the proof of Theorem 4.3.7. For the expansion formula (4.6.12), we define

$$f_{\mu,\nu}(x) := \int_{-\mu}^{\nu} \Phi(x, \lambda) \, d\sigma(\lambda)^T g(\lambda) \qquad (x \in \mathbb{R})$$

4.6. The spectral matrix for the full-line problem

and $h_{cb} := (f - f_{\mu,\nu})\chi_{[0,b]}$, and observe that by isometry

$$(f, h_{cb}) = \int_\mathbb{R} g^T \, d\sigma \, \overline{\mathcal{F}h_b}$$

and

$$\begin{aligned}(f_{\mu,\nu}, h_{cb}) &= \int_\mathbb{R} \left(\int_{-\mu}^\nu \Phi(x,\lambda) \, d\sigma(\lambda)^T \, g(\lambda) \right)^T W(x) \, \overline{h_{cb}(x)} \, dx \\ &= \int_{-\mu}^\nu g(\lambda)^T \, d\sigma(\lambda) \int_\mathbb{R} \Phi(x,\lambda)^T \, W(x) \, \overline{h_{cb}(x)} \, dx \\ &= \int_{-\mu}^\nu g(\lambda)^T \, d\sigma(\lambda) \, \overline{\mathcal{F}h_{cb}(\lambda)}.\end{aligned}$$

Hence the convergence upon letting $b \to \infty$, $c \to -\infty$, and then $\mu, \nu \to \infty$, follows as in the proof of Theorem 4.3.7. □

Denoting, as in section 4.4, the maximal operator of multiplication with the variable λ in $L^2(\mathbb{R}; d\sigma)$ by $\mathcal{M}$, the generalised Fourier transform again gives rise to a self-adjoint operator $H = \mathcal{F}^{-1}\mathcal{M}\mathcal{F}$. Theorem 4.4.3, which establishes the connexion between the spectral properties of this operator and the growth of the spectral function, continues to hold in the present situation, provided that (4.4.5) is interpreted in the sense that the positive semidefinite matrix $\tilde{\sigma} = \sigma(\mu+\epsilon) - \sigma(\mu-\epsilon)$ is not the null matrix.

This condition can be simplified. In fact, $\tilde{\sigma} = 0$ if and only if both $\tilde{\sigma}_{11}$ and $\tilde{\sigma}_{22}$ vanish; indeed, either $\tilde{\sigma}_{12} = 0$ or, since $\tilde{\sigma}_{21}$ and $\tilde{\sigma}_{12}$ are complex conjugates, the positive definiteness of $\tilde{\sigma}$ gives

$$0 \leq \begin{pmatrix} 1 \\ -\tilde{\sigma}_{12}/|\tilde{\sigma}_{12}| \end{pmatrix}^T \begin{pmatrix} \tilde{\sigma}_{11} & \tilde{\sigma}_{12} \\ \tilde{\sigma}_{21} & \tilde{\sigma}_{22} \end{pmatrix} \begin{pmatrix} 1 \\ -\tilde{\sigma}_{21}/|\tilde{\sigma}_{12}| \end{pmatrix} = \tilde{\sigma}_{11} - 2|\tilde{\sigma}_{12}| + \tilde{\sigma}_{22}.$$

Hence in either case

$$|\tilde{\sigma}_{12}| \leq \frac{1}{2}(\tilde{\sigma}_{11} + \tilde{\sigma}_{22}). \tag{4.6.13}$$

This means that the spectral matrix σ is non-constant if and only if at least one of the non-decreasing functions σ_{11} and σ_{22} is non-constant; in other words, the spectrum of H is the union of the sets of growth points of σ_{11} and σ_{22}. In summary, the following analogue of Theorems 4.4.2 and 4.4.3 holds.

Theorem 4.6.3. *The operator H is self-adjoint in $L^2(\mathbb{R} : W)$. A number $\mu \in \mathbb{C}$ is in the spectrum of H if and only if it is real and, for all $\epsilon > 0$,*

$$\sigma(\lambda - \epsilon) - \sigma(\lambda + \epsilon) \neq 0;$$

this is equivalent to

$$\sigma_{11}(\lambda - \epsilon) - \sigma_{11}(\lambda + \epsilon) + \sigma_{22}(\lambda - \epsilon) - \sigma_{22}(\lambda + \epsilon) > 0.$$

Moreover, μ is an eigenvalue if and only if σ is discontinuous at μ.

The proof mirrors that of the statements for the half-line operator. We only mention that, when proving that a discontinuity point of σ is an eigenvalue of H, corresponding to part (e) in the proof of Theorem 4.4.3, the saltus s will be a positive semidefinite non-null matrix and thus have an eigenvector $\xi \in \mathbb{C}^2$ with positive eigenvalue; then take $g := \chi_\mu \xi$ and proceed as before.

The operator H is a realisation of the formal Sturm-Liouville or Dirac differential expressions, as in section 4.4. It has an operator core $\bigcup_{b>0} D_{-b,b}$, where $D_{-b,b}$ is defined as in Lemma 4.6.1 and can be characterised as

$$D_{-b,b} = \{y \in L^2(\mathbb{R}; w) \mid y, py' \in AC_{\text{loc}}(\mathbb{R}), y(x) = 0 \ (x \notin [-b,b])\}$$

for the Sturm-Liouville case and

$$D_{-b,b} = \{u \in L^2(\mathbb{R}; W) \mid u \in AC_{\text{loc}}([0,\infty)), u(x) = 0 \ (x \notin [-b,b])\}$$

in the Dirac case. The operator domains are

$$D(H) = \{y \in L^2(\mathbb{R}; w) \mid y, py' \in AC_{\text{loc}}(\mathbb{R}), \frac{1}{w}(-(py')' + qy) \in L^2(\mathbb{R}; w)\}$$

for the Sturm-Liouville operator and

$$D_{-b,b} = \{u \in L^2(\mathbb{R}; W) \mid u \in AC_{\text{loc}}([0,\infty)),$$
$$-i\sigma_2 u' + (p_1\sigma_3 + p_2\sigma_1 + q)u \in L^2(\mathbb{R}; W)\}$$

for the Dirac operator.

The points in the spectrum of H can be characterised by singular sequences as in the half-line case.

Corollary 4.6.4. *A number $\lambda \in \mathbb{C}$ is in the spectrum of H if and only if there exists a sequence $(f_n)_{n\in\mathbb{N}}$ in $D(H)$ such that $\|f_n\| = 1$ and $\lim_{n\to\infty} \|(H-\lambda)f_n\| = 0$. Without loss of generality, the f_n can be chosen in $\bigcup_{b>0} D_{-b,b}$.*

4.7 The spectrum of the full-line periodic problem

Applying the techniques of the preceding section to the special case of the a-periodic equation considered on the whole real line, we now find that the corresponding self-adjoint operator H has a very simply structured spectrum indeed. It does not have any point spectrum at all, and the spectrum just consists of purely absolutely continuous intervals (called *spectral bands*) which are exactly the closure of the stability intervals of the periodic equation.

Theorem 4.7.1. *The full-line periodic operator H has no eigenvalues. Its spectrum is purely absolutely continuous and equal to the closure of $\mathcal{S}$.*

4.7. The spectrum of the full-line periodic problem

Proof. The first statement follows from the fact that the periodic equation has no non-trivial solutions in $L^2(\mathbb{R}; W)$. Indeed, as we have seen in section 1.4, there is a fundamental system composed of Floquet solutions except possibly at the end-points of the stability intervals $\mathcal{S}$, and these Floquet solutions either have periodic absolute value (case 3) or are exponentially unbounded at ∞ and $-\infty$, respectively (cases 1, 2 and 6). An end-point of $\mathcal{S}$ (cases 4 and 5) is either a point of coexistence, so all solutions are periodic or semi-periodic, or there is a fundamental system composed of a periodic or semi-periodic Floquet solution and a linearly unbounded solution. None of these fundamental systems permits a non-trivial solution square-integrable at both ∞ and $-\infty$.

Therefore the spectral matrix is a continuous function on the real line. It is constant in $\mathcal{I}$, as can be seen in analogy to the proof of Theorem 4.5.1 (b). Indeed, fix $\alpha = \beta \in (0, \pi)$ and consider the boundary-value problem (1.5.1), (4.6.1), (4.6.2) with $c = -na$ and $b = na$, where $n \in \mathbb{N}$. By Theorem 2.4.5 (a), this regular boundary-value problem has at most one eigenvalue in each instability interval, and this eigenvalue is independent of n by periodicity. If $\lambda' < \lambda''$ are such that $[\lambda', \lambda''] \subset \mathcal{I}$ does not contain that eigenvalue, then the spectral matrix is constant in $[\lambda', \lambda'']$ by (4.6.4), and hence so is the spectral matrix σ for the full-line problem, which arises in the limit as $n \to \infty$. Since σ is continuous, σ is therefore constant in each instability interval.

We shall now prove that σ is locally Lipschitz continuous, and hence purely absolutely continuous, in $\mathcal{S}$.

In the following, let $\lambda' < \lambda'' \in \mathbb{R}$ be such that $[\lambda', \lambda''] \subset \mathcal{S}$. Consider the boundary-value problem (1.5.1), (4.6.1), (4.6.2) on $[-na, na]$ as before. By (4.6.4), its spectral matrix $\sigma_n := \sigma_{-na,na,\alpha,\beta}$ satisfies

$$\sigma_{ij}(\lambda'') - \sigma_{ij}(\lambda') = \sum_{\lambda_k \in (\lambda', \lambda'']} \overline{\gamma_k} \gamma_k^T,$$

where $\Phi(\cdot, \lambda_k) \gamma_k$ are the orthonormalised eigenfunctions and λ_k the corresponding eigenvalues.

Furthermore, let $v(\cdot, \lambda)$ be the solution of (1.5.1) with initial condition

$$v(-na, \lambda) = \begin{pmatrix} \sin \alpha \\ \cos \alpha \end{pmatrix}.$$

Then, for each eigenvalue λ_k $v(\cdot, \lambda_k)$ is a multiple of the normalised eigenfunction, and so

$$\frac{1}{\sqrt{\int_{-na}^{na} v(x, \lambda_k)^T W(x) v(x, \lambda_k) \, dx}} v(\cdot, \lambda_k) = \Phi(\cdot, \lambda_k) \gamma_k.$$

Evaluating this identity at the point 0, we obtain

$$\frac{1}{\sqrt{\int_{-na}^{na} v(x, \lambda_k)^T W(x) v(x, \lambda_k) \, dx}} v(0, \lambda_k) = \gamma_k.$$

By Lemma 4.5.2, there exists a constant $C > 0$ such that the Prüfer radius of $v(\cdot, \lambda)$, $\varrho(\cdot, \lambda) = |v(\cdot, \lambda)|$, satisfies

$$\frac{1}{C} \leq |v(x, \lambda)| \leq C$$

for all $x \geq -na$ and all $\lambda \in [\lambda', \lambda'']$. Thus we can estimate

$$|\gamma_k|^2 \leq \frac{C^2}{\int_{-na}^{na} v(x, \lambda_k)^T W(x) v(x, \lambda_k) \, dx} \tag{4.7.1}$$

and

$$|\gamma_k|^2 \geq \frac{1}{C^2 \int_{-na}^{na} v(x, \lambda_k)^T W(x) v(x, \lambda_k) \, dx} \tag{4.7.2}$$

for all eigenvalues $\lambda_k \in [\lambda', \lambda'']$. For the diagonal entries of the spectral matrix σ_n, this yields

$$(\sigma_n)_{ii}(\lambda'') - (\sigma_n)_{ii}(\lambda') = \sum_{\lambda_k \in (\lambda', \lambda'']} |(\gamma_k)_i|^2$$

$$\leq C^2 \sum_{\lambda_k \in (\lambda', \lambda'']} \frac{1}{\int_{-na}^{na} v(x, \lambda_k)^T W(x) v(x, \lambda_k) \, dx}$$

$$= C^2 (\sigma_{2na,\beta}(\lambda'') - \sigma_{2na,\beta}(\lambda')),$$

where $i \in \{1, 2\}$ and $\sigma_{2na,\beta}$ is the spectral function for the regular boundary-value problem on the interval $[0, 2na]$ — which by periodicity is equivalent to our current boundary-value problem — defined in (4.2.14). Passing to the limit $n \to \infty$ and using the estimate (4.5.8) for the half-line spectral function, we thus find that

$$\sigma_{ii}(\lambda'') - \sigma_{ii}(\lambda') \leq \frac{C^4}{\pi}(\lambda'' - \lambda')$$

for $i \in \{1, 2\}$. This shows that the diagonal entries of the full-line spectral matrix are locally Lipschitz continuous in $\mathcal{S}$ and by (4.6.13) this extends to the off-diagonal entries too.

The above estimates also imply that all of $\mathcal{S}$ (and hence also its closure $\overline{\mathcal{S}}$) belongs to the spectrum of H. Indeed, arguing by contradiction, if there were $\lambda' < \lambda''$ such that $[\lambda', \lambda''] \subset \mathcal{S}$ and $\sigma(\lambda'') - \sigma(\lambda') = 0$, then, in view of (4.6.13), for any $\epsilon > 0$ there would be some $n \in \mathbb{N}$ such that

$$(\sigma_n)_{11}(\lambda'') - (\sigma_n)_{11}(\lambda') + (\sigma_n)_{22}(\lambda'') - (\sigma_n)_{22}(\lambda') < \epsilon;$$

but the left-hand side is equal to

$$\sum_{\lambda_k \in (\lambda', \lambda'']} |\gamma_k|^2 \geq \sum_{\lambda_k \in (\lambda', \lambda'']} \frac{1}{C^2 \int_{-na}^{na} v(x, \lambda_k)^T W(x) v(x, \lambda_k) \, dx}$$

$$= \frac{1}{C^2}(\sigma_{2na,\beta}(\lambda'') - \sigma_{2na,\beta}(\lambda')),$$

where we have used (4.7.2) and (4.2.14). Then it would follow in the limit $n \to \infty$ that the spectral function of the half-line problem is constant in $[\lambda', \lambda'']$, contradicting Theorem 4.5.4. □

The simple structure of the spectrum of the periodic operator on the full real line, consisting of the closed stability intervals only, can be used in conjunction with the singular sequence technique of Theorem 4.4.3 (b) to estimate the length of instability intervals, as shown in the following statement.

Corollary 4.7.2. *Let μ be the mid-point of an instability interval of finite length l. Then*
$$l \|f\| \leq 2\|(H - \mu) f\|$$
for any $f \in D(H)$.

Proof. Let $g := \mathcal{F} f \in L^2(\mathbb{R}; d\sigma)$. Then by the Parseval formula,
$$\|(H - \mu) f\|^2 = \int_{\mathbb{R}} |\lambda - \mu|^2 g(\lambda)^T \, d\sigma(\lambda) \, \overline{g(\lambda)}$$
$$= \int_{\mathbb{R} \setminus (\mu - l/2, \mu + l/2)} |\lambda - \mu|^2 g(\lambda)^T \, d\sigma(\lambda) \, \overline{g(\lambda)} \geq \frac{l^2}{4} \|f\|^2,$$
where we have used that σ is constant on the instability interval $(\mu - l/2, \mu + l/2)$ by Theorem 4.7.1. □

This observation has been used in section 3.6 in the following form.

Corollary 4.7.3. *Let μ be the mid-point of an instability interval of finite length l, and $(f_m)_{m \in \mathbb{N}}$ a sequence in $D(H)$ such that $\lim_{m \to \infty} \|f_m\| = 1$. Then*
$$l \leq 2 \liminf_{m \to \infty} \|(H - \mu) f_m\|.$$

4.8 Oscillations and spectra

In sections 4.2, 4.3 and 4.6, we have found a generalised Fourier transform associated with the boundary-value problems on a closed finite interval, on the half-line and on the whole line, assuming the limit-point case at the singular end-points and imposing a boundary condition at each regular end-point. As a consequence, we can characterise spectral subspaces of $D(H)$, where H denotes the corresponding self-adjoint operator, which are invariant under the action of H.

More precisely, if $\Lambda \subset \mathbb{R}$ is an interval (or more generally a σ-measurable subset), then the *spectral subspace* $\mathcal{F}^{-1} L^2(\Lambda; d\sigma)$ is an invariant subspace of H, and its dimension
$$N_\Lambda := \dim \mathcal{F}^{-1} L^2(\Lambda; d\sigma) = \dim L^2(\Lambda; d\sigma)$$
is called the *total multiplicity of the spectrum in* Λ.

Note that we here use the natural embedding of $L^2(\Lambda; d\sigma)$ into $L^2(\mathbb{R}; d\sigma)$, extending elements by 0 outside Λ.

A point $\lambda \in \mathbb{R}$ is called a *point of essential spectrum* if $N_\Lambda = \infty$ for all open intervals $\Lambda \subset \mathbb{R}$ containing λ. Clearly, the regular boundary-value problem considered in section 4.2 has no points of essential spectrum, as its spectrum consists of isolated eigenvalues with a finite-dimensional spectral subspace only. For the spectra of the half-line and full-line operators of Theorems 4.4.2 and 4.6.3, we make the following observation.

Lemma 4.8.1. (a) *If $\lambda \in \mathbb{R}$ lies in the continuous spectrum, then it is an accumulation point of the spectrum.*

(b) *Each accumulation point of the spectrum is a point of essential spectrum.*

Proof. (a) If λ was isolated from the other parts of the spectrum, there would be $\epsilon > 0$ such that λ was the only point of the spectrum in $(\lambda - \epsilon, \lambda + \epsilon)$, contradicting the fact that λ is a point of growth of the continuous part of the spectral function.

(b) If there is a sequence $(\lambda_n)_{n \in \mathbb{N}}$ of points in the spectrum of H with $\lim_{n \to \infty} \lambda_n = \lambda$, then we can find disjoint intervals $\Lambda_n = (\lambda_n - \epsilon_n, \lambda_n + \epsilon_n)$ such that $N_{\Lambda_n} \geq 1$; as any open interval containing λ will include infinitely many of these intervals, the assertion follows. $\square$

Lemma 4.8.1 shows that the continuous spectrum is part of the essential spectrum. Moreover, isolated points of the spectrum are not points of essential spectrum; indeed, by Lemma 4.8.1 (a) they can only be eigenvalues, and then have a finite-dimensional spectral subspace, as any eigenfunction must be a solution of (1.5.1), which has a two-dimensional solution space. In fact, all eigenvalues of the operators H in Lemma 4.8.1 have a one-dimensional eigenspace.

Lemma 4.8.2. *Let $\mu \in \mathbb{R}$, $\delta > 0$ and $\Lambda = (\mu - \delta, \mu + \delta)$. Then N_Λ is the supremum of the dimensions of subspaces $L \subset D(H)$ such that*

$$\|(H - \mu) u\| < \delta \|u\| \qquad (u \in L \setminus \{0\}). \tag{4.8.1}$$

Proof. If $N_\Lambda = \infty$, then setting $\Lambda_\epsilon = (\mu - \delta + \epsilon, \mu + \delta - \epsilon)$ for $0 < \epsilon < \delta$, the dimension of the spaces $L_\epsilon := \mathcal{F}^{-1} L^2(\Lambda_\epsilon; d\sigma)$ is unbounded as $\epsilon \to 0$; by a calculation analogous to (4.4.7), it is clear that (4.8.1) holds for each of these spaces.

Therefore we can assume in the following that $N_\Lambda < \infty$. By Lemma 4.8.1, Λ then only contains finitely many eigenvalues of H and, since Λ is open, there is $\epsilon \in (0, \delta)$ such that $N_{\Lambda_\epsilon} = N_I$. As above, a calculation along the lines of (4.4.7) shows that

$$\|(H - \mu) u\| \leq (\delta - \epsilon) \|u\| < \delta \|u\|$$

for all non-null elements of L_ϵ. Hence this is a subspace of dimension N_Λ which satisfies (4.8.1).

4.8. Oscillations and spectra

Now assume there is a subspace $L \in D(H)$ with $\dim L > N_\Lambda$ and property (4.8.1). Denote by $\lambda_1, \ldots, \lambda_{N_\Lambda}$ the eigenvalues in I. Then the requirement

$$v(\lambda_j) = 0 \qquad (j \in \{1, \ldots, N_\Lambda\})$$

imposes $N_\Lambda < \dim L$ linear conditions on elements $v \in \mathcal{F}L$ and thus will be satisfied on a non-trivial subspace of $\mathcal{F}L$. If $v \neq 0$ is an element of that subspace and $u = \mathcal{F}^{-1}v$, then by the Parseval formula

$$\|(H - \mu)u\|^2 = \int_\mathbb{R} v(\lambda)^T \, d\sigma(\lambda) \, \overline{v(\lambda)}$$

$$= \int_{\mathbb{R} \setminus \Lambda} v(\lambda)^T \, d\sigma(\lambda) \, \overline{v(\lambda)} \geq \delta^2 \|u\|^2,$$

contradicting (4.8.1). $\square$

We conclude this section by proving an analogue of Theorem 2.3.5 which allows approximate counting of eigenvalues of H_α in an interval by studying the oscillation behaviour of solutions of the differential equation at the end-points of that interval. As we use the oscillation results of section 2.3, the following will be valid in the specific situations where H_α is either the Sturm-Liouville or the Dirac operator.

Theorem 4.8.3 (Relative Oscillation Theorem). *Let $\lambda' < \lambda''$, $\alpha \in [0, \pi)$, and $\theta(\cdot, \lambda)$ be the solution of the Prüfer equation (2.2.2) with $\theta(0, \lambda) = \alpha$ ($\lambda \in \{\lambda', \lambda''\}$). Then the total spectral multiplicity of the operator H_α of Theorem 4.4.2 in $(\lambda', \lambda'']$ satisfies*

$$\frac{1}{\pi} \limsup_{b \to \infty}(\theta(b, \lambda'') - \theta(b, \lambda')) - 2 \leq N_{(\lambda', \lambda'']} \leq \frac{1}{\pi} \liminf_{b \to \infty}(\theta(b, \lambda'') - \theta(b, \lambda')) + 2.$$

Proof. Let $n \leq \frac{1}{\pi} \limsup_{b \to \infty}(\theta(b, \lambda'') - \theta(b, \lambda'))$ be a non-negative integer and $\epsilon \in (0, 1)$. Then there is $b > 0$ such that $\frac{1}{\pi}(\theta(b, \lambda'') - \theta(b, \lambda')) > n - \epsilon/2$. Since $\theta(b, \cdot)$ is continuous, there is $\lambda''' \in (\lambda', \lambda'')$ such that $\frac{1}{\pi}(\theta(b, \lambda''') - \theta(b, \lambda')) > n - \epsilon$. Thus, choosing some $\beta \in (0, \pi]$, the regular boundary-value problem (1.5.1), (4.3.2) has at least

$$\frac{1}{\pi}(\theta(b, \lambda''') - \theta(b, \lambda')) - 1 > n - 1 - \epsilon,$$

and so in fact at least $n - 1$, eigenvalues in $(\lambda', \lambda''']$ by Theorem 2.3.5. Let $\lambda_1, \ldots, \lambda_{n-1}$ be these eigenvalues and let $u_1, \ldots, u_{n-1}$ be corresponding orthonormal eigenfunctions. They span an $(n-1)$-dimensional subspace of $D(H)$, all elements of which take a multiple of $\begin{pmatrix} \sin \beta \\ \cos \beta \end{pmatrix}$ as their value at the point b. Hence the linear condition $u(b) = 0$ will be satisfied on an $(n-2)$-dimensional subspace S of this space. The elements of S can be extended by 0 onto $[b, \infty)$, giving elements of $D(H_\alpha)$. When we set $\mu := (\lambda'' + \lambda')/2$ and $\delta := (\lambda'' - \lambda')/2$, Parseval's formula (4.2.8) therefore gives

$$\|(H_\alpha - \mu)u\|^2 = \int_0^b ((H_\alpha - \mu)u)^T W \overline{(H_\alpha - \mu)u} = \sum_{j=1}^{n-1} |\lambda_j - \mu|^2 |(u, u_j)|^2$$

$$< \delta^2 \sum_{j=1}^{n-1} |(u, u_j)|^2 = \delta^2 \|u\|^2 \qquad (u \in S \setminus \{0\}).$$

In conjunction with Lemma 4.8.2, this shows that $N_{(\lambda', \lambda'']} \geq N_{(\lambda', \lambda'')} \geq n - 2$.

For the second inequality, assume that the spectral function σ of H_α has (at least) ν distinct points of increase in (λ', λ''), so that there are points

$$\mu_1 < \lambda_1 < \mu_2 < \lambda_2 < \cdots < \mu_\nu < \lambda_\nu$$

in (λ', λ'') which are not eigenvalues of H_α and such that

$$s_j := \sigma(\lambda_j) - \sigma(\mu_j) > 0 \qquad (j \in \{1, \ldots, \nu\}).$$

By Theorem 4.3.5 (a), there is $b_0 > 0$ such that for any $b > b_0$ the spectral function of the regular boundary-value problem on $[0, b]$ with some boundary condition at b differs by less than $\frac{1}{2} \min_{j \in \{1, \ldots, \nu\}} s_j$ from σ at all points μ_j and λ_j. Thus this boundary-value problem has at least one eigenvalue in each of the disjoint intervals $[\mu_j, \lambda_j]$, and Theorem 2.3.5 shows that

$$\frac{1}{\pi}(\theta(b, \lambda'') - \theta(b, \lambda')) \geq \nu - 1.$$

If (λ', λ'') contains a point of continuous spectrum, Lemma 4.8.1 shows that ν can be taken arbitrarily large. Otherwise (λ', λ'') only contains eigenvalues, and ν can be taken to be the number of eigenvalues, if finite, and arbitrarily large otherwise. Since λ'' may be an additional eigenvalue of H_α, we find that

$$N_{(\lambda', \lambda'']} \leq \frac{1}{\pi} \liminf_{b \to \infty}(\theta(b, \lambda'') - \theta(b, \lambda')) + 2. \qquad \square$$

4.9 Bounded solutions and the absolutely continuous spectrum

The proofs of purely absolutely continuous spectrum for the half-line and full-line operators in the stability intervals, shown in Theorems 4.5.4 and 4.7.1, use the fact that in the case of stability the Floquet solutions are bounded and oscillate in an asymptotically regular way.

More generally, an interval of purely absolutely continuous spectrum arises when all solutions of the differential equation are of comparable size for all values of

4.9. Bounded solutions and the absolutely continuous spectrum

the spectral parameter in this interval. We shall prove this in the present section, focussing on the half-line operator. We do not assume in this section that the coefficients of the equation are periodic.

The main result is the following theorem

Theorem 4.9.1. *Let (1.5.1) be either the Sturm-Liouville or the Dirac system on $[0, \infty)$ with coefficients integrable at 0 and satisfying the general hypotheses. Let $[\lambda', \lambda''] \subset \mathbb{R}$ be an interval with the property that there exists a constant $c > 0$ and a function $k : [0, \infty) \to [0, \infty)$ such that $\lim_{x \to \infty} k(x) = \infty$ and*

$$ck(x) \leq \int_0^x u^T W \overline{u} \leq k(x) \tag{4.9.1}$$

for sufficiently large $x > 0$ and all real-valued solutions u of (1.5.1) with $|u(0)| = 1$.
Then for any $\alpha \in [0, \pi)$ the corresponding self-adjoint operator H_α has purely absolutely continuous spectrum in $[\lambda', \lambda'']$.

In the light of (4.9.1), the hypothesis that k tends to infinity is clearly equivalent to assuming that the limit-point case holds.

In the case of the Sturm-Liouville equation, the condition (4.9.1) takes the form

$$ck(x) \leq \int_0^x |y|^2 w \leq k(x) \tag{4.9.2}$$

for all real-valued solutions y of (1.5.2) with $|y(0)|^2 + |py'(0)|^2 = 1$. For the Dirac system (1.5.4), the condition is simply

$$ck(x) \leq \int_0^x |u|^2 \leq k(x)$$

for all real-valued solutions u with $|u_1(0)|^2 + |u_2(0)|^2 = 1$.

The key to the proof of this theorem is the curious observation that the spectral measures for regular boundary-value problems (1.5.1), (4.3.2), average to Lebesgue measure when we integrate over the boundary condition at 0. We show this for λ intervals; by general measure theoretic considerations the result extends to all measurable sets.

Lemma 4.9.2. *Let $\sigma_{\alpha,b}$ be the spectral function of the regular boundary-value problem (1.5.1), (4.3.2). Then for any interval $[\lambda', \lambda''] \subset \mathbb{R}$,*

$$\int_0^\pi (\sigma_{\alpha,b}(\lambda'') - \sigma_{\alpha,b}(\lambda')) \, d\alpha = \lambda'' - \lambda'. \tag{4.9.3}$$

Proof. Consider the Prüfer variables $\theta(\cdot, \lambda, \alpha), \rho(\cdot, \lambda, \alpha)$ (cf. (2.2.1) of a real-valued solution $v(\cdot, \lambda, \alpha)$ of (1.5.1) for spectral parameter $\lambda \in [\lambda', \lambda'']$, with $\rho(0, \lambda, \alpha) = 1$ and $\theta(0, \lambda, \alpha) = \alpha$.

Differentiating the equation for the Prüfer angle (2.2.2) with respect to the initial value α and using (2.2.3), we find that $\dfrac{\partial \theta}{\partial \alpha}$ satisfies the first-order linear equation

$$\left(\frac{\partial \theta}{\partial \alpha}\right)'(\cdot, \lambda, \alpha) = -2(\log \rho)'(\cdot, \lambda, \alpha) \frac{\partial \theta}{\partial \alpha}(\cdot, \lambda, \alpha)$$

with initial value $\dfrac{\partial \theta}{\partial \alpha}(0, \lambda, \alpha) = 1$, and so

$$\frac{\partial \theta}{\partial \alpha}(x, \lambda, \alpha) = \frac{1}{\rho(x, \lambda, \alpha)^2} \qquad (x > 0).$$

Now we have an eigenvalue for the boundary-value problem exactly if $\beta = \theta(b, \lambda, \alpha)$ (mod π). Differentiating with respect to α and using (4.5.6), we find for the dependence of the eigenvalues on the initial condition α,

$$\frac{\partial \lambda(\alpha)}{\partial \alpha} = -\frac{\partial \theta}{\partial \alpha}(b, \lambda(\alpha), \alpha) \bigg/ \frac{\partial \theta}{\partial \lambda}(b, \lambda(\alpha), \alpha)$$
$$= \frac{-1}{\int_0^b v(x, \lambda(\alpha), \alpha)^T W(x) v(x, \lambda(\alpha), \alpha) \, dx}.$$

Proceeding as in (4.5.5), we conclude that

$$\int_0^\pi (\sigma_{\alpha,b}(\lambda'') - \sigma_{\alpha,b}(\lambda')) \, d\alpha = \int_{\lambda'}^{\lambda''} d\lambda. \qquad \square$$

Assuming the limit-point case at ∞, we can now pass to the limit $b \to \infty$; then $\sigma_{\alpha,b}$ will converge to the spectral function σ_α of the half-line problem with boundary condition (4.3.1) at 0 only. As both the position of the centre and the radius of the circle in Lemma 4.3.2 depend continuously on α, we can see that the constant K in (4.3.20) can be chosen independently of $\alpha \in [0, \pi)$, and so the integrand in (4.9.3) is uniformly bounded. Hence we can apply Lebesgue's dominated convergence theorem and draw the following conclusion.

Theorem 4.9.3. *Assume that the boundary-value problem (1.5.1), (4.3.1) is in the limit-point case, and let σ_α be its spectral function. Then for any interval $[\lambda', \lambda''] \subset \mathbb{R}$,*

$$\int_0^\pi (\sigma_\alpha(\lambda'') - \sigma_\alpha(\lambda')) \, d\alpha = \lambda'' - \lambda'.$$

We now use these observations to prove Theorem 4.9.1.

Proof. For any $\alpha \in [0, \pi)$, $\beta \in (0, \pi]$ and $b > 0$, we can express the difference of the values of the spectral function for the regular boundary-value problem (1.5.1), (4.3.2), between μ and μ', where $\lambda' \leq \mu < \mu' \leq \lambda''$, as in (4.5.4),

$$\sigma_{\alpha,b}(\mu') - \sigma_{\alpha,b}(\mu) = \sum_\lambda \frac{1}{\int_0^b v(x, \lambda, \alpha)^T W(x) v(x, \lambda, \alpha) \, dx},$$

where $v(\cdot, \lambda, \alpha)$ is defined as in the proof of Lemma 4.9.2 and the sum is taken over all eigenvalues λ in $(\mu, \mu']$.

By hypothesis,
$$\int_0^b v(x, \lambda, \alpha)^T W(x) v(x, \lambda, \alpha) \, dx \in [ck(b), k(b)]$$

for all $\lambda \in [\lambda', \lambda'']$ and all $\alpha \in [0, \pi)$; moreover, it is clear from the proof of Theorem 2.3.4 that the number $N_\alpha(\mu, \mu')$ of eigenvalues in $(\mu, \mu']$ for different values of α differs by at most 1. Hence we obtain that for any two $\alpha_1, \alpha_2 \in [0, \pi)$,

$$\sigma_{\alpha_2, b}(\mu') - \sigma_{\alpha_2, b}(\mu) \geq \frac{N_{\alpha_2}(\mu, \mu')}{k(b)}$$

and consequently

$$\sigma_{\alpha_1, b}(\mu') - \sigma_{\alpha_1, b}(\mu) \leq \frac{N_{\alpha_1}(\mu, \mu')}{ck(b)} \leq \frac{1}{c}(\sigma_{\alpha_2, b}(\mu') - \sigma_{\alpha_2, b}(\mu)) + \frac{1}{ck(b)}.$$

Since $\lim_{b \to \infty} k(b) = \infty$, we find in the limit $b \to \infty$ that

$$c(\sigma_{\alpha_1}(\mu') - \sigma_{\alpha_1}(\mu)) \leq \sigma_{\alpha_2}(\mu') - \sigma_{\alpha_2}(\mu).$$

This means that the spectral measures for H_{α_2} gives non-zero weight to any measurable set which has non-zero weight in the spectral measure for H_{α_1}. As α_1 and α_2 are arbitrary, it follows that the spectral measures for H_α for all $\alpha \in [0, \pi)$ give positive weight to the same measurable sets, and in this sense are mutually absolutely continuous.

On the other hand, Theorem 4.9.3 shows that $\int_0^\pi \sigma_\alpha \, d\alpha$ generates Lebesgue measure. If σ_α were not absolutely continuous in $[\lambda', \lambda'']$ for some $\alpha \in [0, \pi)$, then there would be a set $M \subset [\lambda', \lambda'']$ of zero Lebesgue measure such that $\int_M d\sigma_\alpha > 0$. From the above, this would then hold true for all $\alpha \in [0, \pi]$, and integration over α would show that M has positive Lebesgue measure, yielding a contradiction. □

4.10 Chapter notes

§4.2 The Arzelà-Ascoli Theorem can be found in [143, Theorem 15.20]. For the spectral representation theorem for compact symmetric operators (Hilbert-Schmidt Theorem), see [149, Theorem VI.16].

Prüfer [144] gives a beautiful direct proof of Theorem 4.2.3 for the Sturm-Liouville equation.

§4.3 There are different methods of treating the singular Sturm-Liouville or Dirac boundary-value problem. We here follow the traditional approach originating from

Weyl's celebrated paper [198], see also [28, chapter 9] and Titchmarsh's account in [185] and [184]. A rather different approach, based on the theory of operator extensions, is explained in [194]. The spectral function for the Mathieu case was first computed by Eastham, Fulton and Pruess using the software package SLEDGE; see [51] and the references therein.

Helly's selection and integral theorems can be found in e.g. [143, Theorems 12.7 and 12.18, respectively].

Proposition 4.3.3 generalises to $n \times n$ matrix systems, albeit with a somewhat more complicated proof, see [7, Theorem 9.11.2].

§**4.4** In view of Proposition 4.3.4, we discuss the limit-point case only; in the limit-circle case, the situation is briefly as follows. In order to obtain a self-adjoint operator, the domains (4.4.3) and (4.4.4) need to be restricted further by a boundary condition at the singular end-point ∞, requiring the functions u to mimic the behaviour of a chosen solution v in the sense that (in system form) $\lim_{b \to \infty} u(b)^T J \overline{v(b)} = 0$. The resulting operator has a compact resolvent and hence a spectrum consisting of discrete eigenvalues only, and an analogue of Theorem 4.2.3 applies. Details are given in [28].

For Lebesgue's Decomposition Theorem, see [149, Theorem 1.14].

§**4.5** For the Sturm-Liouville case of Theorem 4.5.1 with $\alpha = 0$, see Wallach [191, Theorems I and III]. The technique of averaging over the boundary condition β in (4.5.5) was used in this form in [53]; it is akin to the averaging over the boundary condition α used in the proof of Kotani's Theorem 4.9.3.

The property (simpler than Theorem 4.5.4) that the closure of the stability set constitutes the spectrum goes back to Wintner [200, 201].

We assumed throughout that the weight matrix W is positive semi-definite. Using Krein space instead of Hilbert space techniques, a similar analysis of the periodic Sturm-Liouville operator on the half-line is possible even for indefinite weight w [31].

A similar spectral structure to that of the periodic half-line problem (Theorems 4.5.1 (b) and 4.5.4) appears for the graph Laplacian on a regular tree, where a fixed number (the branching number) of copies of an interval are attached to each right-hand end-point of a copy of that interval. There is a band-gap structure with an eigenvalue of infinite multiplicity between consecutive bands [171, Theorem 3.3]. The infinite multiplicity is due to the high symmetry of the tree and persists when the same locally integrable potential is added to the operator on each tree edge, see [26] (moreover, there is an associated Titchmarsh-Weyl m-function, but it reflects only part of the spectrum, in contrast to the clear link provided by (4.3.16) and (4.4.5) in the half-line case).

§**4.7** In Corollary 4.7.3, the inequality for l in terms of a suitable sequence f_m is not confined to periodic problems. If the condition $f_m \rightharpoonup 0$ is added, the inequality has widespread applications in the spectral theory of differential operators; we refer to Glazman's book [67] and, for example, [50].

4.10. Chapter notes

§4.8 Theorem 4.8.3 and its proof follow [193]. The observation that the difference of Prüfer angles can be used to count eigenvalues even if these angles are individually unbounded has greatly extended the scope of applicability of oscillation methods. For a rather elaborate version of the Relative Oscillation Theorem with a different proof, see [64] for the Sturm-Liouville and [180] for the Dirac system, respectively.

§4.9 The general link between the relative size of the solutions of the differential equation for values of the spectral parameter λ belonging to the support of the point, absolutely continuous or singular continuous spectral measure was discovered by Gilbert and Pearson [66]. Theorem 4.9.1 presents a special case of Gilbert-Pearson subordinacy theory, where the uniform size of all solutions in a λ interval shows that this is an interval of purely absolutely continuous spectrum; this case, known as *uniform non-subordinacy* allows a particularly elegant proof [195] (see also[196]). Theorem 4.9.3 is known as Kotani's theorem [117].

Chapter 5

Perturbations

5.1 Introduction

The periodic Sturm-Liouville or Dirac operator on the whole real line has a purely absolutely continuous spectrum of band-gap structure; the regular end-point of the operator restricted to a half-line only introduces a single eigenvalue, if any, into each spectral gap. In applications, however, one does not always have exact periodicity of the coefficients, and the question arises how the spectral properties of the operator change if a non-periodic perturbation is added to the periodic background potential. In many ways this is analogous to the general question of the spectrum generated by a more or less localised potential added to a free Sturm-Liouville or Dirac operator, but here we take as an unperturbed reference a periodic operator, whose spectral properties are very well known by the results shown in the preceding chapters.

We begin by noting in section 5.2 that the spectral bands remain intervals of purely absolutely continuous spectrum under a very mild decay condition on the perturbation. In section 5.3, we observe that if the perturbation tends to 0 at infinity, then every compact subinterval of an instability interval contains at most finitely many eigenvalues and no further spectrum. In particular, this means that the instability intervals, while not devoid of spectrum in general, continue to be gaps in the essential spectrum. We also derive asymptotics for the distribution of eigenvalues thus introduced into the gaps in the limit of slow variation (sometimes called the adiabatic limit) of the perturbation. The question of whether an instability interval as a whole contains a finite or infinite number of eigenvalues turns out to have a more subtle answer, given in section 5.4. There is a critical boundary case for perturbations with x^{-2} asymptotic decay, and the critical coupling constant can be expressed in terms of the derivative of Hill's discriminant at the point of transition between instability and stability. In the supercritical case, where eigenvalues in the gap accumulate at a band edge, we find their asymptotic distribution in section 5.5, showing that they are exponentially close to the band.

5.2 Spectral bands

We have seen in section 4.3 that the periodic Sturm-Liouville and Dirac operators on the half-line have purely absolutely continuous spectrum in the set $\mathcal{S}$ of stability intervals of the corresponding periodic differential equation system. In the present section it is shown that this property is stable when the coefficients of the operator are perturbed by the addition of non-periodic terms which satisfy a mild decay condition at infinity, stipulating essentially that their local average tends to zero and their oscillations can be controlled.

For a 2×2 matrix S, we denote by $|S|$ the matrix operator norm,

$$|S| = \sup_{v \in \mathbb{C}^2 \setminus \{0\}} \frac{|Sv|}{|v|}.$$

Then for two matrices S_1 and S_2, we have $|S_1 S_2| \leq |S_1| |S_2|$. Moreover, convergence of a matrix sequence in the norm sense implies convergence for each entry separately.

Theorem 5.2.1. *Let B and W be 2×2 matrix-valued functions on $[0, \infty)$ satisfying the general hypotheses of section 1.5, and assume that $B = B_1 + B_2$, where B_1 and W are a-periodic and B_2 has the properties*

$$\int_a^\infty |B_2(t) - B_2(t-a)| \, dt < \infty, \tag{5.2.1}$$

$$\lim_{x \to \infty} \int_x^{x+a} |B_2| = 0. \tag{5.2.2}$$

Let $[\lambda', \lambda''] \subset \mathcal{S}$, where $\mathcal{S}$ is the stability set of the periodic equation

$$u' = J(B_1 + \lambda W) u. \tag{5.2.3}$$

Then there is a constant $C > 0$ such that $|u(x, \lambda)| < C$ for all $\lambda \in [\lambda', \lambda'']$ and all solutions $u(\cdot, \lambda)$ of

$$u' = J(B_1 + B_2 + \lambda W) u \tag{5.2.4}$$

such that $|u(0, \lambda)| = 1$.

Proof. Let Φ and Ψ be the canonical fundamental matrices of the periodic equation (5.2.3) and of the perturbed periodic equation (5.2.4), respectively. For $j \in \mathbb{N}$, let Ψ_j be the solution of (5.2.4) with initial value $\Psi_j(a(j-1)) = I$; we then set $M_j := \Psi_j(aj)$. Then $\Phi_j(x) := \Phi(x - a(j-1))$ will serve an analogous purpose for the unperturbed equation (5.2.3), with $\Phi_j(aj) = M$, the monodromy matrix of (5.2.3), for all j.

5.2. Spectral bands

Rewriting (5.2.4) in the form

$$u' = J(B_1 + \lambda W)\, u + J B_2\, u,$$

we find by the variation of constants formula (1.2.11) that

$$\Psi_j(x) = \Phi_j(x) + \Phi_j(x) \int_{a(j-1)}^{x} \Phi_j^{-1}\, J\, B_2\, \Psi_j \qquad (x \geq a(j-1)). \tag{5.2.5}$$

Denoting in the following by (const.) a uniform constant for all $\lambda \in [\lambda', \lambda'']$ — although not always the same constant — we find from (5.2.5) that

$$|\Psi_j(x)| \leq |\Phi_j(x)| \left(1 + \int_{a(j-1)}^{x} |\Phi_j^{-1} J|\, |B_2|\, |\Psi_j| \right),$$

and hence by Gronwall's lemma the estimate

$$|\Psi_j(x)| \leq (\text{const.}) \exp\left((\text{const.}) \int_{a(j-1)}^{x} |B_2| \right), \tag{5.2.6}$$

for $x \in [a(j-1), aj]$. Using this in combination with (5.2.5) again, we obtain for such x,

$$|\Psi_j(x) - \Phi_j(x)| \leq \int_{a(j-1)}^{x} |\Phi_j(x)\, \Phi_j^{-1} J|\, |B_2|\, |\Psi_j|$$

$$\leq (\text{const.}) \int_{a(j-1)}^{x} |B_2(t)| \exp\left((\text{const.}) \int_{a(j-1)}^{t} |B_2| \right) dt$$

$$\leq (\text{const.}) \left(\int_{a(j-1)}^{aj} |B_2| \right) \exp\left((\text{const.}) \int_{a(j-1)}^{aj} |B_2| \right)$$

$$\to 0 \qquad (j \to \infty) \tag{5.2.7}$$

because of (5.2.2). In particular, taking $x = aj$, we find that

$$|M_j - M| = |\Psi_j(aj) - \Phi_j(aj)| \to 0 \quad (j \to \infty) \tag{5.2.8}$$

uniformly for $\lambda \in [\lambda', \lambda'']$. Setting $D_j = \operatorname{Tr} M_j$, we conclude that $\lim_{j \to \infty} |D_j - D| = 0$, where D is the discriminant of (5.2.3). Since $[\lambda', \lambda''] \subset \mathcal{S}$, this implies that, for a sufficiently large $J \in \mathbb{N}$ and some $\delta > 0$,

$$|D_j(\lambda)| \leq 2 - \delta$$

for all $j > J$ and $\lambda \in [\lambda', \lambda'']$.

The matrix M_j has determinant 1 and hence can be analysed as in our study of the monodromy matrix in section 1.4. For $j > J$, we are in Case 3; so M_j has complex conjugate eigenvalues $\mu_j, \overline{\mu_j}$ with $|\mu_j| = 1$ and corresponding eigenvalues given in terms of the eigenvalues and the entries of M_j by a formula analogous to (4.5.2). In view of the convergence of M_j to the monodromy matrix M in (5.2.8), these eigenvectors converge to those of M, uniformly in $[\lambda', \lambda'']$, as $j \to \infty$.

Let E_j be the matrix of eigenvectors (4.5.2) for M_j, $j > J$. Then E_j converges to the matrix E of eigenvectors of M, and so E_j^{-1} converges to E^{-1} and hence is bounded, uniformly in $\lambda \in [\lambda', \lambda'']$. From

$$M_j = E_j \begin{pmatrix} \mu_j & 0 \\ 0 & \overline{\mu_j} \end{pmatrix} E_j^{-1}$$

we obtain

$$\Psi(na) = M_n M_{n-1} \cdots M_{J+1} \Psi(Ja)$$
$$= E_n \begin{pmatrix} \mu_n & 0 \\ 0 & \overline{\mu_n} \end{pmatrix} E_n^{-1} E_{n-1} \begin{pmatrix} \mu_{n-1} & 0 \\ 0 & \overline{\mu_{n-1}} \end{pmatrix} E_{n-1}^{-1} \cdots$$
$$\cdots E_{J+1} \begin{pmatrix} \mu_{J+1} & 0 \\ 0 & \overline{\mu_{J+1}} \end{pmatrix} E_{J+1}^{-1} \Psi(Ja)$$

and, since $\begin{pmatrix} \mu_j & 0 \\ 0 & \overline{\mu_j} \end{pmatrix}$ is unitary, it follows that

$$|\Psi(na)| \le |E_n| |E_n^{-1} E_{n-1}| |E_{n-1}^{-1} E_{n-2}| \cdots |E_{J+2}^{-1} E_{J+1}| |E_{J+1}^{-1}| |\Psi(Ja)|. \quad (5.2.9)$$

In order to estimate $|E_j^{-1} E_{j-1}|$, we observe that $\Psi_{j-1}(\cdot - a)$ is a fundamental matrix of

$$u'(x) = J(B_1(x) + B_2(x-a) + \lambda W(x)) u(x),$$

and so by the variation of constants formula (1.2.11)

$$\Psi_j(x) = \Psi_{j-1}(x-a)$$
$$+ \Psi_{j-1}(x-a) \int_{a(j-1)}^{x} \Psi_{j-1}(t-a)^{-1} J (B_2(t) - B_2(t-a)) \Psi_j(t) \, dt.$$

Using (5.2.6) and the convergence of Ψ_{j-1} to Φ_{j-1}, we can therefore estimate

$$|M_j - M_{j-1}| = |\Psi_j(aj) - \Psi_{j-1}(a(j-1))|$$
$$\le (\text{const.}) \int_{a(j-1)}^{aj} |B_2(t) - B_2(t-a)| \, dt$$

and, in view of (4.5.2), also

$$|E_j - E_{j-1}| \le (\text{const.}) \int_{a(j-1)}^{aj} |B_2(t) - B_2(t-a)| \, dt.$$

5.2. Spectral bands

Hence, observing that
$$|E_j^{-1}E_j| = |I - E_j^{-1}(E_j - E_{j-1})| \leq 1 + |E_j^{-1}||E_j - E_{j-1}|,$$
we can follow up on (5.2.9),

$$|\Psi(na)| \leq |E_n||E_{J+1}^{-1}||\Psi(Ja)| \prod_{j=J+2}^{n} |E_j^{-1}E_{j-1}|$$

$$\leq (\text{const.}) \prod_{j=J+2^n} \left(1 + (\text{const.}) \int_{a(j-1)}^{aj} |B_2(t) - B_2(t-a)|\, dt\right)$$

$$\leq (\text{const.}) \exp\left((\text{const.}) \sum_{j=J+2}^{n} \int_{a(j-1)}^{aj} |B_2(t) - B_2(t-a)|\, dt\right)$$

$$\leq (\text{const.}) \exp\left((\text{const.}) \int_{a(j+1)}^{\infty} |B_2(t) - B_2(t-a)|\, dt\right) < \infty.$$

In conjunction with the uniform boundedness of Ψ_n, this shows that Ψ, and hence any solution $u(\cdot, \lambda)$ of (5.2.4) with $|u(0, \lambda)| = 1$, is bounded uniformly with respect to $\lambda \in [\lambda', \lambda'']$. □

We remark that the condition (5.2.2) can, in a sense, already be inferred from (5.2.1). Indeed, if we consider the shifted functions
$$B_{2,n}(x) := B_2(x + na) \quad (x \in [0, a]; n \in \mathbb{N}),$$
then we find that, for $n, m \in \mathbb{N}$ with $m < n$,

$$\int_0^a |B_{2,n} - B_{2,m}| = \int_0^a \left|\sum_{j=m}^{n-1}(B_2(x + (j+1)a)) - B_2(x + ja))\right| dx$$

$$\leq \sum_{j=m}^{n-1} \int_0^a |B_2(x + (j+1)a) - B_2(x + ja)|\, dx$$

$$= \int_{ma}^{na} |B_2(x + a) - B_2(x)|\, dx \to 0 \quad (m, n \to \infty)$$

by (5.2.1), which shows that $B_{2,n}$ converges to a limit $B_{2,\infty}$ in $L_1([0, a])$. We extend $B_{2,\infty}$ to an a-periodic function on $[0, \infty)$. Then $B = \tilde{B}_1 + \tilde{B}_2$, where $\tilde{B}_1 := B_1 + B_{2,\infty}$ and $\tilde{B}_2 := B_2 - B_{2,\infty}$ satisfy both (5.2.1) and (5.2.2). Note, however, that now $\mathcal{S}$ in Theorem 5.2.1 will be the stability set for the periodic equation
$$u' = J(\tilde{B}_1 + \lambda W)u.$$
Theorems 5.2.1 and 4.9.1 give the following statement which shows that the absolutely continuous spectral bands of the periodic Dirac operator (see Theorem 4.5.4) are preserved under perturbations satisfying a mild decay condition.

Corollary 5.2.2. Let p_1, p_2 and q be locally integrable, a-periodic real-valued functions on $[0, \infty)$. Moreover, let $\tilde{p}_1, \tilde{p}_2$ and $\tilde{q}$ be locally integrable real-valued functions satisfying

$$\int_a^\infty |\tilde{p}_1(t) - \tilde{p}_1(t-a)|\, dt < \infty, \qquad \lim_{x \to \infty} \int_x^{x+a} |\tilde{p}_1| = 0$$

(and similarly for $\tilde{p}_2, \tilde{q}$). Then, for any $\alpha \in [0, \pi)$ the one-dimensional Dirac operator

$$H_\alpha = -i\sigma_2 \frac{d}{dx} + (p_1 + \tilde{p}_1)\sigma_3 + (p_2 + \tilde{p}_2)\sigma_1 + (q + \tilde{q})$$

with boundary condition (4.3.1) has purely absolutely continuous spectrum in the stability set $\mathcal{S}$ of the periodic Dirac equation (1.5.4).

Proof. Let $[\lambda', \lambda''] \subset \mathcal{S}$. Then by Theorem 5.2.1 there exists a constant C such that for all $\lambda \in [\lambda', \lambda'']$, all solutions of the perturbed periodic equation

$$-i\sigma_2 u' + ((p_1 + \tilde{p}_1)\sigma_3 + (p_2 + \tilde{p}_2)\sigma_1 + q + \tilde{q})\, u = \lambda u$$

with $|u(0)| = 1$ are bounded: $|u(x)| < C$ $(x \geq 0)$. By the same reasoning as at the end of the proof of Lemma 4.5.2, this also implies that $|u(x)| > 1/C$ $(x \geq 0)$. In particular,

$$\frac{k(x)}{C^4} \leq \int_0^x |u|^2 \leq k(x) \qquad (x \geq 0)$$

with $k(x) := C^2 x$. Theorem 4.9.1 now shows that $[\lambda', \lambda'']$ is an interval of purely absolutely continuous spectrum of H_α. $\square$

For the perturbed Hill equation, the required lower bound on the growth of the square-integral of solutions y is slightly more difficult to obtain. Nevertheless, we have the following analogue of Corollary 5.2.2.

Corollary 5.2.3. Let $p > 0$, $w > 0$ and q be locally integrable, a-periodic real-valued functions on $[0, \infty)$. Moreover, let $\tilde{p}$ and $\tilde{q}$ be locally integrable real-valued functions such that $p + \tilde{p} > 0$,

$$\int_a^\infty |\tilde{q}(t) - \tilde{q}(t-a)|\, dt < \infty, \qquad \lim_{x \to \infty} \int_x^{x+a} |\tilde{q}| = 0$$

and

$$\int_a^\infty \frac{1}{p(t)} \left| \frac{\tilde{p}}{p+\tilde{p}}(t) - \frac{\tilde{p}}{p+\tilde{p}}(t-a) \right| dt < \infty, \qquad \lim_{x \to \infty} \int_x^{x+a} \left| \frac{\tilde{p}}{p(p+\tilde{p})} \right| = 0.$$

Then, for any $\alpha \in [0, \pi)$ the one-dimensional Sturm-Liouville operator

$$H_\alpha = \frac{1}{w}\left(-\frac{d}{dx}((p+\tilde{p})\frac{d}{dx}) + (q+\tilde{q}) \right)$$

5.2. Spectral bands

with boundary condition

$$y(0)\cos\alpha - (py')(0)\sin\alpha = 0$$

has purely absolutely continuous spectrum in the stability set $\mathcal{S}$ of the periodic Sturm-Liouville equation (1.5.2).

Proof. Let $[\lambda', \lambda''] \subset \mathcal{S}$. Then, as in the proof of Corollary 5.2.2, we can use Theorem 5.2.1 to find a constant $C > 0$ such that, for all $\lambda \in [\lambda', \lambda'']$, all solutions of the perturbed periodic Sturm-Liouville system

$$u'(x,\lambda) = \begin{pmatrix} 0 & \frac{1}{p+\tilde{p}} \\ q+\tilde{q}-\lambda w & 0 \end{pmatrix} u(x,\lambda) \tag{5.2.10}$$

with $|u(0,\lambda)| = 1$ satisfy $1/C \leq |u(x,\lambda)| \leq C$ for all $x \geq 0$.

Let y be a real-valued solution of the perturbed Sturm-Liouville equation

$$-((p+\tilde{p})\,y')' + (q+\tilde{q})\,y = \lambda\,w\,y$$

such that $|y(0)|^2 + |(py')(0)|^2 = 1$; then $u = \begin{pmatrix} y \\ py' \end{pmatrix}$ will be a solution of (5.2.10) with the required property. Now for $n \in \mathbb{N}$, let Ψ_n and Φ_n be defined as in the proof of Theorem 5.2.1. Then, for $x \in [(n-1)a, na]$,

$$u(x) = \Psi_n(x)\,u((n-1)a)$$
$$= \Phi_n(x)\,u((n-1)a) + (\Psi_n(x) - \Phi_n(x))\,u((n-1)a)$$

and therefore by Minkowski's inequality

$$\sqrt{\int_{(n-1)a}^{na} |y|^2\,w} \geq \sqrt{\int_{(n-1)a}^{na} |[\Phi_n\,u((n-1)a)]_1|^2}$$
$$- \sqrt{\int_{(n-1)a}^{na} |[(\Psi_n - \Phi_n)\,u((n-1)a)]_1|^2}.$$

As $|u((n-1)a)| < C$ for all n and $\Psi_n - \Phi_n \to 0$ uniformly on the interval of integration as $n \to \infty$ by (5.2.7), the last term tends to 0 in this limit. For the first term on the right-hand side, we observe that $[\Phi_n\,u((n-1)a)]_1$ is a real-valued solution of the periodic equation with

$$|\Phi_n((n-1)a)\,u((n-1)a)| = |u((n-1)a)| \geq 1/C,$$

and so by Lemma 4.5.3 we have

$$\int_{(n-1)a}^{na} |[\Phi_n\,u((n-1)a)]_1|^2\,w \geq \frac{C'}{C^2}$$

with a constant C' which only depends on $[\lambda', \lambda'']$.

Hence we see that, for sufficiently large $x > 0$, (4.9.2) will be satisfied with $k(x) = C^2 x$ and $c = C'/2C^4$. The assertion now follows by Theorem 4.9.1. □

5.3 Gap eigenvalues

We now turn to the instability intervals of the periodic system. As we have seen in section 4.5, the essential spectrum of the unperturbed periodic operator on the half-line has gaps coinciding with the instability set $\mathcal{I}$, with each gap containing either no spectrum at all or only a single eigenvalue. We shall now show that, when a perturbation is added which tends to 0 at ∞, the qualitative picture remains unchanged; indeed, each instability interval contains only discrete eigenvalues and thus is still a gap in the essential spectrum. Every compact subinterval of an instability interval contains at most a finite number of eigenvalues. However, the question whether the whole instability interval contains only finitely many eigenvalues, or eigenvalues which accumulate at one or both of its end-points, is more subtle and will be considered in section 5.4.

Regarding the number of eigenvalues in a given subinterval of an instability interval, we then observe that, in the adiabatic or homogenisation limit where the perturbation (which is assumed to be continuous) varies on a very long scale compared to the period, the number of eigenvalues generally increases asymptotically linearly in the scaling parameter and has a limit density which can be conveniently expressed in terms of the rotation number of the periodic equation.

Theorem 5.3.1. *Let $[\lambda', \lambda''] \subset \mathcal{I}$, where $\mathcal{I}$ is the instability set of Hill's equation (1.5.3). Moreover, let $\alpha \in [0, \pi)$ and let $\tilde{q}$ be a locally integrable, real-valued function on $[0, \infty)$ such that*

$$\lim_{x \to \infty} \frac{\tilde{q}(x)}{w(x)} = 0.$$

Then the perturbed periodic Sturm-Liouville operator

$$H_\alpha = \frac{1}{w}\left(-\frac{d}{dx}(p\frac{d}{dx}) + q + \tilde{q}\right)$$

has at most finitely many eigenvalues and no other spectrum in $[\lambda', \lambda'']$.

Proof. As $\mathcal{I}$ is open, there exists $\delta > 0$ such that $[\lambda' - \delta, \lambda'' + \delta] \subset \mathcal{I}$. Let x_0 be an integer multiple of a such that

$$\frac{|\tilde{q}(x)|}{w(x)} \leq \delta \qquad (x \geq x_0),$$

and set

$$q_\pm(x) := \begin{cases} q(x) + \tilde{q} & \text{if } x \in [0, x_0), \\ q(x) \mp \delta w(x) & \text{if } x \in [x_0, \infty). \end{cases}$$

Now for $\lambda \in \{\lambda', \lambda''\}$, let $\theta(x, \lambda)$ ($x \geq 0$) be the solution of the initial-value problem for the Prüfer equation

$$\theta'(x, \lambda) = \frac{1}{p(x)}\cos^2\theta(x, \lambda) + (\lambda w(x) - q(x) - \tilde{q}(x))\sin^2\theta(x, \lambda), \quad \theta(0, \lambda) = \alpha,$$
(5.3.1)

5.3. Gap eigenvalues

and similarly $\theta_\pm(x, \lambda)$ ($x \geq 0$) the solutions of

$$\theta'_\pm(x, \lambda) = \frac{1}{p(x)} \cos^2 \theta_\pm(x, \lambda) + (\lambda w(x) - q_\pm(x)) \sin^2 \theta_\pm(x, \lambda), \quad \theta_\pm(0, \lambda) = \alpha. \tag{5.3.2}$$

Then, since $q_+ \leq q + \tilde{q} \leq q_+$ throughout, comparison of (5.3.1) with (5.3.2) and Theorem 2.3.1 (a) show that

$$\theta_-(x, \lambda) \leq \theta(x, \lambda) \leq \theta_+(x, \lambda) \tag{5.3.3}$$

for all $x \geq 0$; the three functions are identical on $[0, x_0]$.

Let $n \in \mathbb{Z}$ be the index, according to the enumeration of Theorem 2.4.1, of the instability interval in which both $\lambda' - \delta$ and $\lambda'' + \delta$ lie. Then, noting that for $x \geq x_0$ the coefficient of the last term in (5.3.2) is

$$\lambda w(x) - q_\pm(x) = (\lambda \pm \delta) w(x) - q(x)$$

and thus (5.3.2) is the Prüfer equation for the periodic equation (1.5.3) with spectral parameter $\lambda \pm \delta$, we can apply (2.4.1) to find that

$$\theta_+(x, \lambda'') = \theta(x_0, \lambda'') + \frac{n\pi}{a}(x - x_0) + O(1),$$

$$\theta_-(x, \lambda') = \theta(x_0, \lambda') + \frac{n\pi}{a}(x - x_0) + O(1)$$

asymptotically for $x \to \infty$. Hence, using (5.3.3), we conclude that

$$\theta(x, \lambda'') - \theta(x, \lambda') \leq \theta_+(x, \lambda'') - \theta_-(x, \lambda')$$
$$= \theta(x_0, \lambda'') - \theta(x_0, \lambda') + O(1);$$

in particular, the difference remains bounded as $x \to \infty$. The finiteness of the total spectral multiplicity of H_α in $[\lambda', \lambda'']$ now follows by Theorem 4.8.3. $\square$

In the case of the Dirac operator, the following analogue holds for general matrix-valued perturbations.

Theorem 5.3.2. *Let $[\lambda', \lambda''] \subset \mathcal{I}$, where $\mathcal{I}$ is the instability set of the periodic Dirac equation (1.5.4). Moreover, let $\alpha \in [0, \pi)$, and let $\tilde{p}_1, \tilde{p}_2$ and $\tilde{q}$ be locally integrable, real-valued functions on $[0, \infty)$ such that*

$$\lim_{x \to \infty} \tilde{p}_1(x) = \lim_{x \to \infty} \tilde{p}_2(x) = \lim_{x \to \infty} \tilde{q}(x) = 0. \tag{5.3.4}$$

Then the perturbed periodic Dirac operator

$$H_\alpha = -i\sigma_2 \frac{d}{dx} + (p_1 + \tilde{p}_1)\sigma_3 + (p_2 + \tilde{p}_2)\sigma_1 + (q + \tilde{q})$$

has at most finitely many eigenvalues and no other spectrum in $[\lambda', \lambda'']$.

Proof. We proceed in analogy to the proof of Theorem 5.3.1. Again, there is $\delta > 0$ such that $[\lambda' - \delta, \lambda'' + \delta] \subset \mathcal{I}$. The Prüfer equation (2.2.2) for the perturbed Dirac system takes the form

$$\theta' = \begin{pmatrix} \sin\theta \\ \cos\theta \end{pmatrix}^T \left(B(x) + \tilde{B}(x) + \lambda I \right) \begin{pmatrix} \sin\theta \\ \cos\theta \end{pmatrix},$$

where

$$B = \begin{pmatrix} -p_1 - q & -p_2 \\ -p_2 & p_1 - q \end{pmatrix}, \qquad \tilde{B} = \begin{pmatrix} -\tilde{p}_1 - \tilde{q} & -\tilde{p}_2 \\ -\tilde{p}_2 & \tilde{p}_1 - \tilde{q} \end{pmatrix}.$$

Hypothesis (5.3.4) ensures that there is $x_0 > 0$ such that the pointwise operator norm of the perturbation matrix $\tilde{B}$ satisfies $|\tilde{B}(x)| \le \delta$ for all $x \ge x_0$. Consequently, for such x the matrices

$$\delta I \pm \tilde{B}(x)$$

are positive semidefinite. Hence

$$\begin{pmatrix} \sin\theta \\ \cos\theta \end{pmatrix}^T (B(x) + (\lambda - \delta)I) \begin{pmatrix} \sin\theta \\ \cos\theta \end{pmatrix} \le \begin{pmatrix} \sin\theta \\ \cos\theta \end{pmatrix}^T \left(B(x) + \tilde{B}(x) + \lambda I \right) \begin{pmatrix} \sin\theta \\ \cos\theta \end{pmatrix}$$

$$\le \begin{pmatrix} \sin\theta \\ \cos\theta \end{pmatrix}^T (B(x) + (\lambda + \delta)I) \begin{pmatrix} \sin\theta \\ \cos\theta \end{pmatrix}$$

for any θ and $x \ge x_0$. Theorem 2.3.1 (a) then implies that the Prüfer angle θ of the perturbed equation with initial value $\theta(0) = \alpha$ can be estimated above and below by the Prüfer angles $\theta_\pm$ of the equation where the perturbation matrix $\tilde{B}$ is replaced with the constant matrix $\pm \delta I$ on $[x_0, \infty)$, in analogy to (5.3.3). The remainder of the proof is exactly as for Theorem 5.3.1. $\square$

The key idea of Theorems 5.3.1 and 5.3.2 is to use the monotonicity of Prüfer angles under perturbations to eventually replace the perturbation with a constant and then apply the growth asymptotic of the Prüfer angle for the periodic equation, as obtained in section 2.4. The same idea can be adapted to estimating how many eigenvalues appear in any subinterval of an instability interval under the influence of a continuous perturbation in the limit of slow variation. More precisely, given a continuous function $\tilde{q}$ which serves as a template, we consider perturbations of the form $\tilde{q}(x/c)$, where c is a dilation parameter which tends to infinity in the limit. Clearly, the local modulus of continuity of the perturbation decreases towards zero as c increases, which means that the perturbation changes ever more slowly on the length scale defined by the period a. This limit is related to the adiabatic limit in quantum mechanics, which refers to perturbations changing slowly in time compared to the dynamic time scale of the unperturbed system, and to the homogenisation limit, in which microscopic material properties, here represented by the periodic background, are treated by averaging in contrast to the macroscopic structures.

5.3. Gap eigenvalues

Specifically for the perturbed periodic Sturm-Liouville equation we have the following result.

Theorem 5.3.3. *Let $[\lambda', \lambda''] \subset \mathcal{I}$, where $\mathcal{I}$ is the instability set of Hill's equation (1.5.2) with $w = 1$. Moreover, let $\alpha \in [0, \pi)$ and let $\tilde{q}$ be a continuous real-valued function on $[0, \infty)$ with $\lim_{r \to \infty} \tilde{q}(r) = 0$.*

Then the number of eigenvalues in $[\lambda', \lambda'']$ of the perturbed periodic Sturm-Liouville operator

$$H_\alpha = -\frac{d}{dx}\left(p\frac{d}{dx}\right) + q(x) + \tilde{q}(x/c)$$

has asymptotic

$$N_{[\lambda',\lambda'']} \sim \frac{c}{\pi a} \int_0^\infty \Big(k(\lambda'' - \tilde{q}(r)) - k(\lambda' - \tilde{q}(r))\Big)\, dr \qquad (c \to \infty), \qquad (5.3.5)$$

where k is the rotation number of the unperturbed equation (1.5.2).

Proof. Let $\delta > 0$ be so small that $[\lambda' - \delta, \lambda'' + \delta] \subset \mathcal{I}$. Then there is $r_0 > 0$ such that $|\tilde{q}(r)| \le \delta$ for all $r \ge r_0$. Let $\theta(\cdot, \lambda)$ be the solution of the initial-value problem (5.3.1) for the perturbed Prüfer equation, with spectral parameter $\lambda \in \{\lambda', \lambda''\}$.

Now let $m \in \mathbb{N}$ and consider a dissection of the interval $[0, r_0]$ into m parts, i.e. division points $0 = s_0 < s_1 < \cdots < s_m = r_0$. For $j \in \{1, \ldots, m\}$, let

$$\tilde{q}_j^- = \sup_{s \in [s_{j-1}, s_j]} \tilde{q}(s), \qquad \tilde{q}_j^+ = \inf_{s \in [s_{j-1}, s_j]} \tilde{q}(s),$$

and let $\theta_j^\pm(\cdot, \lambda)$ be the solutions of the Prüfer equations

$$(\theta_j^\pm)'(x, \lambda) = \frac{1}{p(x)} \cos^2 \theta_j^\pm(x, \lambda) + (\lambda - \tilde{q}_j^\pm(x) - q(x)) \sin^2 \theta_j^\pm(x, \lambda), \qquad (5.3.6)$$

with initial condition $\theta_j^\pm(cs_{j-1}, \lambda) = \theta(cs_{j-1}, \lambda)$. Then by Sturm comparison (Corollary 2.3.2), we find that

$$\theta_j^-(cs_j, \lambda) - \theta_j^-(cs_{j-1}, \lambda) \le \theta(cs_j, \lambda) - \theta(cs_{j-1}, \lambda) \le \theta_j^+(cs_j, \lambda) - \theta_j^+(cs_{j-1}, \lambda). \qquad (5.3.7)$$

On the other hand, the equations (5.3.6) have a-periodic coefficients and are in fact the Prüfer equations for the unperturbed periodic Sturm-Liouville equation with spectral parameter shifted by $-\tilde{q}_j^\pm$. Therefore the asymptotics (2.4.9) (with $k = n\pi$ in the instability interval I_n) apply, giving

$$\theta_j^\pm(cs_j, \lambda) - \theta_j^\pm(cs_{j-1}, \lambda) = k(\lambda - \tilde{q}_j^\pm) \frac{cs_j - cs_{j-1}}{a} + O(1) \qquad (5.3.8)$$

in the limit $c \to \infty$.

Also, by the same reasoning as in the proof of Theorem 5.3.1, we find that for $\lambda \in \{\lambda', \lambda''\}$ and $x > cr_0$,

$$\theta(x, \lambda) - \theta(cr_0, \lambda) = \frac{n\pi}{a}(x - cr_0) + O(1),$$

the remainder staying bounded as $x \to \infty$, where n is the number of the instability interval such that $[\lambda', \lambda''] \subset I_n$; we here use the fact that $\lambda' - \delta, \lambda'' + \delta \in I_n$.

Now appealing to the Relative Oscillation Theorem 4.8.3, we find that the number of eigenvalues of H_α in $(\lambda', \lambda'']$ has the asymptotic

$$\lim_{c \to \infty} \frac{N_{(\lambda', \lambda'']}}{c} = \lim_{c \to \infty} \frac{1}{\pi c} \left(\theta(cr_0, \lambda'') - \theta(cr_0, \lambda') \right)$$

$$= \lim_{c \to \infty} \frac{1}{\pi c} \sum_{j=1}^{m} \Big((\theta(cs_j, \lambda'') - \theta(cs_{j-1}, \lambda'')) - (\theta(cs_j, \lambda') - \theta(cs_{j-1}, \lambda')) \Big).$$

Hence, using the estimates (5.3.7) and the asymptotics (5.3.8), we conclude that

$$\frac{1}{\pi a} \sum_{j=1}^{m} \left(k(\lambda'' - \tilde{q}_j^-) - k(\lambda' - \tilde{q}_j^+) \right) (s_j - s_{j-1})$$

$$\leq \lim_{c \to \infty} \frac{N_{(\lambda', \lambda'']}}{c} \leq \frac{1}{\pi a} \sum_{j=1}^{m} \left(k(\lambda'' - \tilde{q}_j^+) - k(\lambda' - \tilde{q}_j^-) \right) (s_j - s_{j-1}).$$

The statement of Theorem 5.3.3 now follows by observing that the sums on either side are lower and upper Riemann sums corresponding to the given dissection for the integral

$$\int_0^{r_0} \left(k(\lambda'' - \tilde{q}(r)) - k(\lambda' - \tilde{q}(r)) \right) dr$$

and that $k(\lambda'' - \tilde{q}(r)) = k(\lambda' - \tilde{q}(r)) = n\pi$ if $r \geq r_0$. □

For the perturbed periodic Dirac system, the situation is a bit more complicated, mostly due to the fact that perturbations often apply to the matrix coefficients p_1 and p_2 in practice, so the simple estimate using upper and lower Riemann sums, as used in the proof of Theorem 5.3.3 above, needs to be replaced with less tight matrix operator norm estimates. For example, the angular momentum term arising from the separation in spherical polar coordinates of a three-dimensional radially periodic Dirac operator has the form $\sigma_1 k/r$, with r the radial variable, and thus can be considered a perturbation of p_2. The angular momentum term is also singular at 0. In the the following we shall focus on problems with one regular end-point; see the notes for the doubly singular case.

Moreover, the integral for the asymptotic density of eigenvalues will also involve values of the rotation number of the periodic equation where not only the spectral parameter, but also the coefficients p_1 and p_2 are shifted by a constant. Specifically, we shall denote by $k(\lambda, c_1, c_2)$ the rotation number of the periodic Dirac equation

$$-i\sigma_2 u' + p_1 \sigma_3 u + p_2 \sigma_1 u + q u = (\lambda - c_1 \sigma_3 - c_2 \sigma_1) u, \qquad (5.3.9)$$

where p_1, p_2, q are real-valued, locally integrable and a-periodic, and $\lambda, c_1, c_2 \in \mathbb{R}$. As

$$c_1 \sigma_3 + c_2 \sigma_1 + (|c_1| + |c_2|) I \geq 0, \qquad (|c_1| + |c_2|) I - c_1 \sigma_3 - c_2 \sigma_1 \geq 0$$

5.3. Gap eigenvalues

in the sense of positive semidefinite matrices, Sturm comparison (Corollary 2.3.2) and the continuity of the rotation number as a function of the spectral parameter show that k is jointly continuous in all three variables.

Theorem 5.3.4. *Let $[\lambda', \lambda''] \subset \mathcal{I}$, where $\mathcal{I}$ is the instability set of the periodic Dirac equation (1.5.4). Let $\alpha \in [0, \pi)$ and let $\tilde{q}, \tilde{p}_1, \tilde{p}_2$ be continuous real-valued functions on $[0, \infty)$ with $\lim_{r \to \infty} \tilde{q}(r) = \lim_{r \to \infty} \tilde{p}_1(r) = \lim_{r \to \infty} \tilde{p}_2(r) = 0$.*

Then the number of eigenvalues in $[\lambda', \lambda'']$ of the perturbed periodic Dirac operator

$$H_\alpha = -i\sigma_2 \frac{d}{dx} + (p_1(x) + \tilde{p}_1(x/c))\sigma_3 + (p_2(x) + \tilde{p}_2(x/c))\sigma_1 + q(x) + \tilde{q}(x/c)$$

has asymptotic

$$N_{[\lambda', \lambda'']} \sim \frac{c}{\pi a} \int_0^\infty \Big(k(\lambda'' - \tilde{q}(r), \tilde{p}_1(r), \tilde{p}_2(r)) - k(\lambda' - \tilde{q}(r), \tilde{p}_1(r), \tilde{p}_2(r))\Big) dr \quad (5.3.10)$$

as $c \to \infty$.

Proof. Let $\delta > 0$ be so small that $\lambda'' > \lambda' + 2\delta$ and $[\lambda' - \delta, \lambda'' + \delta] \in \mathcal{I}$. Then there is $r_0 > 0$ such that $|\tilde{q}(r)| + |\tilde{p}_1(r)| + |\tilde{p}_2(r)| < \delta$ for $r \geq r_0$. Moreover, there is a bound $M > 0$ such that $|\tilde{q}(r)|, |\tilde{p}_1(r)|, |\tilde{p}_2(r)| \leq M$ for all $r \geq 0$.

Let $\epsilon > 0$. As k is uniformly continuous on

$$K = [\lambda' - \delta - M, \lambda'' + \delta + M] \times [-M, M]^2,$$

there is $\tilde{\delta} \in (0, \delta]$ such that, for $(\lambda, c_1, c_2), (\tilde{\lambda}, \tilde{c}_1, \tilde{c}_2) \in K$,

$$|\lambda - \tilde{\lambda}|, |c_1 - \tilde{c}_1|, |c_2 - \tilde{c}_2| < \tilde{\delta} \Rightarrow |k(\lambda, c_1, c_2) - k(\tilde{\lambda}, \tilde{c}_1, \tilde{c}_2)| < \epsilon.$$

Since $\tilde{p}_1, \tilde{p}_2$ and $\tilde{q}$ are uniformly continuous on $[0, r_0]$, there exists $\gamma > 0$ such that, for $x, y \in [0, r_0]$,

$$|x - y| < \gamma \Rightarrow |\tilde{q}(x) - \tilde{q}(y)|, |\tilde{p}_1(x) - \tilde{p}_1(y)|, |\tilde{p}_2(x) - \tilde{p}_2(y)| < \tilde{\delta}/3.$$

Now consider a dissection of $[0, r_0]$ into m subintervals with dissection points $0 = s_0 < s_1 < \cdots < s_m = r_0$ such that $|s_l - s_{j-1}| < \gamma$. Choose $\hat{s}_j \in [s_{j-1}, s_j]$ and set

$$c_{1,j} = \tilde{p}_1(\hat{s}_j), \qquad c_{2,j} = \tilde{p}_2(\hat{s}_j), \qquad c_{3,j} = \tilde{q}(\hat{s}_j),$$

for each $j \in \{1, \ldots m\}$. Then on $[s_{j-1}, s_j]$,

$$(\tilde{p}_1 - c_{1,j})\sigma_3 + (\tilde{p}_2 - c_{2,j})\sigma_1 + \tilde{q} - c_{3,j} + \tilde{\delta} I \geq 0,$$
$$\tilde{\delta} I - (\tilde{p}_1 - c_{1,j})\sigma_3 - (\tilde{p}_2 - c_{2,j})\sigma_1 - (\tilde{q} - c_{3,j}) \geq 0 \quad (5.3.11)$$

in the sense of positive semidefinite matrices. Let $\theta(\cdot, \lambda)$ be the solution of the Prüfer equation for the perturbed periodic Dirac equation (cf. (2.2.7)),

$$\theta'(x, \lambda) = \lambda - q(x) - \tilde{q}(x/c) + (p_1(x) + \tilde{p}_1(x/c)) \cos 2\theta(x, \lambda)$$
$$- (p_2(x) + \tilde{p}_2(x/c)) \sin 2\theta(x, \lambda)$$

with initial value $\theta(0, \lambda) = \alpha$. For $j \in \{1, \ldots, m\}$, let $\tilde{\theta}_{j,k}(\cdot, \lambda)$ with $k \in \{1, 2\}$ be the solutions on $[cs_{j-1}, cs_j]$ of the periodic Prüfer equation corresponding to (5.3.9),

$$\theta'_{j,k}(x, \lambda) = \lambda - q(x) - c_{3,j} + (p_1(x) + c_{1,j}) \cos 2\theta_{j,k}(x, \lambda)$$
$$- (p_2(x) + c_{2,j}) \sin 2\theta_{j,k}(x, \lambda)$$

with initial values

$$\theta_{j,1}(cs_{j-1}, \lambda) = \theta(cs_{j-1}, \lambda'), \qquad \theta_{j,2}(cs_{j-1}, \lambda) = \theta(cs_{j-1}, \lambda'').$$

Then, using the estimates (5.3.11) in Sturm comparison (Corollary 2.3.2), we find that

$$\tilde{\theta}_{j,1}(cs_j, \lambda' - \tilde{\delta}) - \tilde{\theta}_{j,1}(cs_{j-1}, \lambda' - \tilde{\delta}) \leq \theta(cs_j, \lambda') - \theta(cs_{j-1}, \lambda')$$
$$\leq \tilde{\theta}_{j,1}(cs_j, \lambda' + \tilde{\delta}) - \tilde{\theta}_{j,1}(cs_{j-1}, \lambda' + \tilde{\delta}),$$
$$\tilde{\theta}_{j,2}(cs_j, \lambda'' - \tilde{\delta}) - \tilde{\theta}_{j,2}(cs_{j-1}, \lambda'' - \tilde{\delta}) \leq \theta(cs_j, \lambda'') - \theta(cs_{j-1}, \lambda'')$$
$$\leq \tilde{\theta}_{j,2}(cs_j, \lambda'' + \tilde{\delta}) - \tilde{\theta}_{j,2}(cs_{j-1}, \lambda'' + \tilde{\delta}).$$

From (2.4.9) we have for $\mu \in \mathbb{R}$,

$$\tilde{\theta}_{j,k}(cs_j, \mu) - \tilde{\theta}_{j,k}(sc_{j-1}, \mu) = k(\mu - c_{3,j}, c_{1,j}, c_{2,j}) \frac{c(s_j - s_{j-1})}{a} + O(1)$$

in the limit $c \to \infty$. Therefore we find that

$$\lim_{c \to \infty} \frac{1}{\pi c} (\theta(cr_0, \lambda'') - \theta(cr_0, \lambda'))$$
$$\geq \lim_{c \to \infty} \frac{1}{\pi c} \sum_{j=1}^{m} \Big((\tilde{\theta}_{j,2}(cs_j, \lambda'' - \tilde{\delta}) - \tilde{\theta}_{j,2}(cs_{j-1}, \lambda'' - \tilde{\delta}))$$
$$- (\tilde{\theta}_{j,1}(cs_j, \lambda' + \tilde{\delta}) - \tilde{\theta}_{j,1}(cs_{j-1}, \lambda' + \tilde{\delta})) \Big)$$
$$= \frac{1}{\pi a} \sum_{j=1}^{m} (s_j - s_{j-1}) \Big(k(\lambda'' - \tilde{\delta} - c_{3,j}, c_{1,j}, c_{2,j}) - k(\lambda' + \tilde{\delta} - c_{3,j}, c_{1,j}, c_{2,j}) \Big)$$
$$\geq \frac{1}{\pi a} \sum_{j=1}^{m} (s_j - s_{j-1}) \Big(k(\lambda'' - c_{3,j}, c_{1,j}, c_{2,j}) - k(\lambda' - c_{3,j}, c_{1,j}, c_{2,j}) - 2\epsilon \Big)$$

and analogously

$$\lim_{c\to\infty} \frac{1}{\pi c}(\theta(cr_0,\lambda'') - \theta(cr_0,\lambda'))$$

$$\leq \lim_{c\to\infty} \frac{1}{\pi c}\sum_{j=1}^{m}\Big((\tilde{\theta}_{j,2}(cs_j,\lambda''+\tilde{\delta}) - \tilde{\theta}_{j,2}(cs_{j-1},\lambda''+\tilde{\delta}))$$

$$- (\tilde{\theta}_{j,1}(cs_j,\lambda'-\tilde{\delta}) - \tilde{\theta}_{j,1}(cs_{j-1},\lambda'-\tilde{\delta}))\Big)$$

$$= \frac{1}{\pi a}\sum_{j=1}^{m}(s_j - s_{j-1})\Big(k(\lambda''+\tilde{\delta} - c_{3,j},c_{1,j},c_{2,j}) - k(\lambda'-\tilde{\delta} - c_{3,j},c_{1,j},c_{2,j})\Big)$$

$$\leq \frac{1}{\pi a}\sum_{j=1}^{m}(s_j - s_{j-1})\Big(k(\lambda'' - c_{3,j},c_{1,j},c_{2,j}) - k(\lambda' - c_{3,j},c_{1,j},c_{2,j}) + 2\epsilon\Big).$$

As in the proof of Theorem 5.3.3, the Prüfer angles remain bounded independently of c on $[cr_0,\infty)$. Furthermore, the above Riemann sums converge to the corresponding integrals due to the uniform continuity of the integrand, and the integrand of (5.3.10) vanishes on $[cr_0,\infty)$. Thus, the Relative Oscillation Theorem 4.8.3 gives

$$\frac{1}{\pi a}\int_0^\infty (k(\lambda'' - \tilde{q}(r),\tilde{p}_1(r),\tilde{p}_2(r)) - k(\lambda' - \tilde{q}(r),\tilde{p}_1(r),\tilde{p}_2(r)))\, dr - \frac{2\epsilon r_0}{\pi a}$$

$$\leq \lim_{c\to\infty} \frac{N_{(\lambda',\lambda'']}}{c}$$

$$\leq \frac{1}{\pi a}\int_0^\infty (k(\lambda'' - \tilde{q}(r),\tilde{p}_1(r),\tilde{p}_2(r)) - k(\lambda' - \tilde{q}(r),\tilde{p}_1(r),\tilde{p}_2(r)))\, dr + \frac{2\epsilon r_0}{\pi a}.$$

As $\epsilon > 0$ was arbitrary, the statement of Theorem 5.3.4 follows. □

5.4 Critical coupling constants

We now turn to the question whether a perturbed periodic Sturm-Liouville or Dirac operator has a finite or infinite total number of eigenvalues in an instability interval of the unperturbed periodic equation. The results of section 5.3 have shown that any compact subinterval of an instability interval contains at most finitely many eigenvalues. Therefore it only remains to settle the question whether or not eigenvalues accumulate at an end-point of the instability interval. The answer depends on the rate of decay of the perturbation. We shall see in the following that the critical decay rate is x^{-2}, and that the exact value of the asymptotic constant at this scale is crucial.

We begin by considering the relative oscillation of a real-valued solution of the perturbed periodic equation

$$w' = J(B + \tilde{B} + \lambda W)w \tag{5.4.1}$$

compared to a real-valued solution of the periodic equation (1.5.1). At the moment, we only assume that the perturbation $\tilde{B}$ is a locally integrable, real 2×2 matrix-valued function, but there will be further restrictions later. By the general assumption that B and W are symmetric, $\operatorname{Tr}(J(B+\lambda W)) = 0$ throughout; so we can consider linearly independent, $\mathbb{R}^2$-valued solutions u and v of (1.5.1) whose Wronskian $W(u,v) = 1$. Let $\Psi = (u,v)$ be the fundamental matrix formed from these solutions.

Then we combine the idea of variation of constants (cf. Proposition 1.2.2) with that of the Prüfer transformation (2.2.1), writing

$$w = \Psi \hat{a} \begin{pmatrix} \sin\gamma \\ \cos\gamma \end{pmatrix} \tag{5.4.2}$$

with a non-zero amplitude function $\hat{a}$ and a relative angle function γ; from (1.5.1) and (5.4.1) we obtain

$$J \tilde{B} \Psi \hat{a} \begin{pmatrix} \sin\gamma \\ \cos\gamma \end{pmatrix} = \Psi \hat{a}' \begin{pmatrix} \sin\gamma \\ \cos\gamma \end{pmatrix} + \Psi \hat{a} \gamma' \begin{pmatrix} \cos\gamma \\ -\sin\gamma \end{pmatrix}.$$

Multiplying from the left with $\begin{pmatrix} \cos\gamma \\ -\sin\gamma \end{pmatrix}^T$, this gives

$$\gamma' = \begin{pmatrix} \cos\gamma \\ -\sin\gamma \end{pmatrix}^T \Psi^{-1} J \tilde{B} \Psi \begin{pmatrix} \sin\gamma \\ \cos\gamma \end{pmatrix} = \begin{pmatrix} \sin\gamma \\ \cos\gamma \end{pmatrix}^T \Psi^T \tilde{B} \Psi \begin{pmatrix} \sin\gamma \\ \cos\gamma \end{pmatrix}; \tag{5.4.3}$$

in the last step we used the identity

$$\Psi^{-1} = -J\Psi^T J \tag{5.4.4}$$

which is easily verified for any 2×2 matrix of determinant 1 by direct calculation.

The relative angle variable γ serves as a suitable proxy for the difference between the Prüfer angles of w and of the solution u of the unperturbed equation, as the next lemma shows.

Lemma 5.4.1. *Let $u,v : [0,\infty) \to \mathbb{R}^2$ be solutions of (1.5.1) with Wronskian $W(u,v) = 1$, and let $w : [0,\infty) \to \mathbb{R}^2$ be a non-trivial solution of (5.4.1). Let θ and θ_1 be Prüfer angles of w and u, respectively, and let γ be defined as in (5.4.2) and such that $\theta(0) - \theta_1(0)$ and $\gamma(0) - \frac{\pi}{2}$ lie in the same interval $[n\pi, (n+1)\pi]$ with $n \in \mathbb{Z}$.*

Then $|\gamma - (\theta - \theta_1 + \frac{\pi}{2})| < \pi$.

Proof. Let R, R_1, R_2 be the Prüfer radii of w, u and v and θ_2 a Prüfer angle of v. Then (5.4.2) can be rewritten as

$$A \begin{pmatrix} \sin\gamma \\ \cos\gamma \end{pmatrix} = R \begin{pmatrix} R_2 \cos\theta_2 & -R_2 \sin\theta_2 \\ -R_1 \cos\theta_1 & R_1 \sin\theta_1 \end{pmatrix} \begin{pmatrix} \sin\theta \\ \cos\theta \end{pmatrix} = R \begin{pmatrix} R_2 \sin(\theta-\theta_2) \\ -R_1 \sin(\theta-\theta_1) \end{pmatrix},$$

5.4. Critical coupling constants

which gives

$$\tan \gamma = -\frac{R_2}{R_1} \frac{\sin(\theta - \theta_1 + \theta_1 - \theta_2)}{\sin(\theta - \theta_1)}$$
$$= \frac{R_2}{R_1} \sin(\theta_1 - \theta_2) \left(\tan(\theta - \theta_1 + \frac{\pi}{2}) - \cot(\theta_1 - \theta_2) \right).$$

Now considering that $1 = W(u,v) = R_1 R_2 \sin(\theta_1 - \theta_2)$, we see that the first two factors on the right-hand side are positive, and that $\cot(\theta_1 - \theta_2)$ is locally absolutely continuous. Therefore γ is related to $\theta - \theta_1 + \frac{\pi}{2}$ by a Kepler transformation, and the assertion follows by Theorem 2.2.1. □

In the following we assume that, for $x \geq 1$, the perturbation is of the specific form

$$\tilde{B}(x) = \frac{1}{x^2} (\hat{B} + \beta(x)), \qquad \beta(x) = o(1) \quad (x \to \infty), \tag{5.4.5}$$

with a constant real symmetric matrix $\hat{B}$. As the only condition on β is that it tends to 0 at infinity, this assumption is purely asymptotic and does not impose any restrictions, beyond the general hypotheses, on $\tilde{B}$ in any compact interval.

Moreover, we assume that in (1.5.1) λ is an end-point of an instability interval, u is a corresponding periodic or semi-periodic solution, and we take v to be the solution arising from u by Rofe-Beketov's formula

$$v = fJu + gu \tag{5.4.6}$$

with real-valued functions f, g (cf. Theorem 1.9.1). As we are studying the question whether γ is unbounded or not, and γ is continuous, it is clearly sufficient to consider the differential equation (5.4.3) for γ on the interval $[1, \infty)$, where it takes the form

$$\gamma' = \frac{1}{x^2} (u \sin \gamma + gu \cos \gamma - fJu \cos \gamma)^T (\hat{B} + \beta(x))$$
$$\times (u \sin \gamma + gu \cos \gamma + fJu \cos \gamma)$$
$$= \cos^2 \gamma \left(\frac{1}{x} (\tan \gamma + g) u + \frac{1}{x} fJu \right)^T (\hat{B} + \beta(x)) \left(\frac{1}{x} (\tan \gamma + g) u + \frac{1}{x} fJu \right). \tag{5.4.7}$$

As g is locally absolutely continuous, we can perform the Kepler transformation

$$\tan \phi(x) = \frac{1}{x} (\tan \gamma(x) + g(x)),$$

whereupon (2.2.12) gives the differential equation for ϕ,

$$\phi' = \frac{1}{x} \Big(-\sin \phi \cos \phi + g' \cos^2 \phi$$
$$+ \cos^2 \phi (u \tan \phi + \frac{1}{x} fJu)^T (\hat{B} + \beta(x)) (u \tan \phi + \frac{1}{x} fJu) \Big)$$

$$= \frac{1}{x}\left(-\sin\phi\cos\phi + g'\cos^2\phi + \sin^2\phi\, u^T(\hat{B}+\beta(x))u\right)$$
$$+ \frac{1}{x^2}\sin\phi\cos\phi\, f\left((Ju)^T(\hat{B}+\beta(x))u + u^T(\hat{B}+\beta(x))Ju\right)$$
$$+ \frac{1}{x^3}\cos^2\phi\, f^2(Ju)^T(\hat{B}+\beta(x))(Ju).$$

If we introduce the a-periodic functions
$$F_1 := g', \qquad F_2 := u^T\hat{B}u, \qquad G := u^T\beta u, \qquad (5.4.8)$$

and use the fact that $f = -|u|^{-2}$ (see (1.9.8)) is a-periodic and therefore bounded, we can rewrite the above differential equation for ϕ more briefly in the form

$$\phi' = \frac{1}{x}\left(F_1\cos^2\phi - \sin\phi\cos\phi + (F_2+G)\sin^2\phi\right) + O(x^{-2}) \qquad (x\to\infty). \tag{5.4.9}$$

Since ϕ and γ are connected by a Kepler transformation, ϕ is as good an indicator as γ of the asymptotic boundedness or otherwise of the difference of Prüfer angles of w and u. We now observe that F_1 and F_2 are a-periodic and therefore the analysis of the differential equation (5.4.9) can be much simplified by averaging ϕ over a period interval.

Lemma 5.4.2. *Let $F_1, F_2 : [0,\infty) \to \mathbb{R}$ be locally integrable and a-periodic, $G : [1,\infty) \to \mathbb{R}$ locally integrable with $\lim_{x\to\infty} G(x) = 0$ and $\phi : [1,\infty) \to \mathbb{R}$ a locally absolutely continuous function such that (5.4.9) holds. Then the averaged function*

$$\tilde{\phi}(x) := \frac{1}{a}\int_x^{x+a}\phi \qquad (x\geq 1)$$

is locally absolutely continuous, $\lim_{x\to\infty}|\phi(x) - \tilde{\phi}(x)| = 0$, and

$$\tilde{\phi}'(x) = \frac{1}{x}\left(C_1\cos^2\tilde{\phi} - \sin\tilde{\phi}\cos\tilde{\phi} + \left(C_2 + \frac{1}{a}\int_x^{x+a}G\right)\sin^2\tilde{\phi}\right) + O(x^{-2})$$

as $x\to\infty$, where $C_j = \frac{1}{a}\int_0^a F_j$, $j\in\{1,2\}$.

Proof. By the Mean Value Theorem for integrals, for each $x\geq 1$ there is an $x' \in [x, x+a]$ such that $\tilde{\phi}(x) = \phi(x')$. Hence for all $t\in[x, x+a]$,

$$|\phi(t) - \tilde{\phi}(x)| = \left|\int_{x'}^t \phi'\right| \leq \frac{1}{x}\int_0^a\left(|F_1| + \frac{1}{2} + |F_2|\right) + o(x^{-1}) = O(x^{-1}) \tag{5.4.10}$$

$(x\to\infty)$, and in particular $\lim_{x\to\infty}|\phi(x) - \tilde{\phi}(x)| = 0$.

5.4. Critical coupling constants

Clearly $\tilde{\phi}$ is locally absolutely continuous, and using (5.4.9) and integrating by parts we obtain

$$\tilde{\phi}'(x) = \frac{1}{a} \int_x^{x+a} \phi'$$

$$= -\frac{1}{a} \left[\frac{1}{t} \int_t^{x+a} (F_1 \cos^2 \phi - \sin \phi \cos \phi + (F_2 + G) \sin^2 \phi) \right]\Big|_x^{x+a}$$

$$\quad - \frac{1}{a} \int_x^{x+a} \frac{1}{t^2} \int_t^{x+a} (F_1 \cos^2 \phi - \sin \phi \cos \phi + (F_2 + G) \sin^2 \phi) \, dt + O\left(\frac{1}{x^2}\right)$$

$$= \frac{1}{ax} \int_x^{x+a} (F_1 \cos^2 \phi - \sin \phi \cos \phi + (F_2 + G) \sin^2 \phi) + O(x^{-2}).$$

In view of

$$\left.\begin{array}{l} |\sin^2 z_1 - \sin^2 z_2| \\ |\cos^2 z_1 - \cos^2 z_2| \\ |\sin z_1 \cos z_1 - \sin z_2 \cos z_2| \end{array}\right\} \leq |z_1 - z_2| \quad (z_1, z_2 \in \mathbb{R}) \quad (5.4.11)$$

and the estimate (5.4.10), ϕ can be replaced with $\tilde{\phi}$ in the integral while keeping the same asymptotic order for the remainder term. $\square$

The differential equation for $\tilde{\phi}$ in Lemma 5.4.2 has asymptotically constant coefficients. This makes it possible to find a simple criterion to decide whether or not its solutions are globally bounded.

Lemma 5.4.3. *Let $C_1, C_2 \in \mathbb{R}$, and let $h : [1, \infty) \to \mathbb{R}$ be locally integrable with $h(x) = o(x^{-1})$ $(x \to \infty)$. Let $\tilde{\phi} : [1, \infty) \to \mathbb{R}$ be a locally absolutely continuous function such that*

$$\tilde{\phi}'(x) = \frac{1}{x}(C_1 \cos^2 \tilde{\phi}(x) - \sin \tilde{\phi}(x) \cos \tilde{\phi}(x) + C_2 \sin^2 \tilde{\phi}(x)) + h(x) \quad (5.4.12)$$

$(x \geq 1)$. Then $\tilde{\phi}$ is bounded if $C_1 C_2 < 1/4$ and unbounded if $C_1 C_2 > 1/4$.

Proof. Choosing the constant $\phi_0 \in \mathbb{R}$ such that

$$\sin 2\phi_0 = \frac{-1}{\sqrt{1 + (C_1 - C_2)^2}}, \quad \cos 2\phi_0 = \frac{C_1 - C_2}{\sqrt{1 + (C_1 - C_2)^2}},$$

we can rewrite

$$C_1 \cos^2 \tilde{\phi} - \sin \tilde{\phi} \cos \tilde{\phi} + C_2 \sin^2 \tilde{\phi} = \frac{C_1 + C_2}{2} + \frac{\sqrt{1 + (C_1 - C_2)^2}}{2} \cos 2(\tilde{\phi} - \phi_0).$$

Then the function $\psi = \tilde{\phi} - \phi_0$ satisfies

$$\psi'(r) = \frac{1}{2x}\left(C_1 + C_2 + \sqrt{1 + (C_1 - C_2)^2} \cos 2(\tilde{\phi}(x) - \phi_0)\right) + h(x) \quad (x \geq 1). \quad (5.4.13)$$

By the hypothesis on h, and since we assume that $C_1 C_2 \neq 1/4$, there exists a point $x_0 > 1$ such that

$$|4xh(x)| \leq \left|\sqrt{1 + (C_1 - C_2)^2} - |C_1 + C_2|\right|$$

for all $x \geq x_0$.

Now assume that $C_1 C_2 < 1/4$, which is equivalent to

$$|C_1 + C_2| < \sqrt{1 + (C_1 - C_2)^2}.$$

Then for $x \geq x_0$, the right-hand side of (5.4.13) is strictly positive if $\psi = 0 \pmod{\pi}$ and strictly negative if $\psi = \frac{\pi}{2} \pmod{\pi}$, and hence $\psi(x)$ is trapped in the interval $(n\pi, n\pi + \frac{3\pi}{2})$, where $n \in \mathbb{Z}$ is such that $\psi(x_0)$ lies in this interval. Hence ψ, and consequently $\tilde{\phi}$, are globally bounded.

In the case $C_1 C_2 > 1/4$, which is equivalent to $|C_1 + C_2| > \sqrt{1 + (C_1 - C_2)^2}$,

$$|\psi(x) - \psi(x_0)| \geq \frac{1}{4}\left(|C_1 + C_2| - \sqrt{1 + (C_1 - C_2)^2}\right)\log\frac{x}{x_0} \to \infty \qquad (x \to \infty),$$

and so ψ and $\tilde{\phi}$ are unbounded. □

The critical product $C_1 C_2$ can be conveniently expressed in terms of the properties of the discriminant of the periodic equation. Let $\tilde{D}(c)$ be the discriminant of the system

$$u' = J\left(B + c\hat{B} + \lambda W\right)u, \qquad (5.4.14)$$

where $\hat{B}$ is the constant matrix of (5.4.5) and we assume as before that λ is an end-point of an instability interval of the unperturbed equation. Thus $\tilde{D}(0) = \pm 2$. Then we have the following.

Lemma 5.4.4. *The constants C_1 and C_2 of Lemma 5.4.2 satisfy*

$$C_1 C_2 = -\frac{1}{a^2}|\tilde{D}|'(0).$$

Proof. Let $\Phi(x, c)$ be the canonical fundamental matrix of (5.4.14), so that

$$\Phi'(x, c) = J\left(B(x) + c\hat{B} + \lambda W(x)\right)\Phi(x, c), \qquad \Phi(0, c) = I.$$

Then $\frac{\partial}{\partial c}\Phi(x, 0)$ is the solution of the initial-value problem

$$\frac{\partial \Phi'}{\partial c}(x, 0) = J\left(B(x) + \lambda W(x)\right)\frac{\partial \Phi}{\partial c}(x, 0) + J\hat{B}\Phi(x, 0), \qquad \frac{\partial \Phi}{\partial c}(0, 0) = 0.$$

Using the variation of constants formula (1.2.11), (1.2.15) and $\tilde{D}(c) = \operatorname{Tr}\Phi(a, c)$, we hence find

$$\frac{\partial \tilde{D}}{\partial c}(0) = \operatorname{Tr}\left(\Phi(a) J \int_0^a \Phi^T(s) \hat{B} \Phi(s)\, ds\right), \qquad (5.4.15)$$

5.4. Critical coupling constants

where $\Phi = \Phi(\cdot, 0)$.

Now let $\Psi = (u, v)$ be the fundamental matrix of the unperturbed periodic equation, as considered above in (5.4.2), with a-periodic or a-semi-periodic u and v as in (5.4.6). Then $\Psi(x) = \Phi(x)\Psi(0)$ and therefore

$$\begin{aligned}\operatorname{Tr}(\Phi(a)\,\Phi(s)^{-1}\,J\,\hat{B}\,\Phi(s)) &= \operatorname{Tr}(\Psi(a)\,\Psi(s)^{-1}\,J\,\hat{B}\,\Psi(s)\,\Psi(0)^{-1}) \\ &= \operatorname{Tr}(\Psi(0)^{-1}\,\Psi(a)\,J\,\Psi(s)^T\,\hat{B}\,\Psi(s)),\end{aligned} \qquad (5.4.16)$$

bearing in mind identity (5.4.4) and the fact that the trace of a product of matrices is invariant under cyclic permutation. Since $u(a) = \pm u(0)$,

$$\Psi(0)^{-1}\,\Psi(a) = \begin{pmatrix} \pm 1 & v_2(0)v_1(a) - v_1(0)v_2(a) \\ 0 & \pm 1 \end{pmatrix}. \qquad (5.4.17)$$

Further, $f(a) = f(0) = |u(0)|^{-2}$ and so

$$v(0) = f(0)\,J\,u(0), \qquad v(a) = \pm(f(0)\,J\,u(0) + g(a)\,u(0)),$$

which together with (5.4.17) gives

$$\Psi(0)^{-1}\,\Psi(a) = \pm \begin{pmatrix} 1 & g(a) \\ 0 & 1 \end{pmatrix}. \qquad (5.4.18)$$

Also

$$\Psi^T\,\hat{B}\,\Psi = \begin{pmatrix} u^T\hat{B}u & u^T\hat{B}v \\ v^T\hat{B}u & v^T\hat{B}v \end{pmatrix}. \qquad (5.4.19)$$

Taking (5.4.15), (5.4.16), (5.4.18), (5.4.19) and (5.4.8) together, we arrive at

$$\frac{\partial \tilde{D}}{\partial c}(0) = \mp g(a)\int_0^a u^T\,\hat{B}\,u = \mp \int_0^a F_1 \int_0^a F_2,$$

and the assertion follows because $|\tilde{D}(c)| = \pm\tilde{D}(c)$ for c in a neighbourhood of 0. $\square$

These considerations give the following criterion for the finiteness of the number of gap eigenvalues for the perturbed Hill's equation.

Theorem 5.4.5. *Let $\lambda^{(n)}$ be an end-point of an instability interval I_n of Hill's equation (1.5.2) with $w = 1$. Moreover, let $\alpha \in [0, \pi)$ and let $\tilde{q}$ be a locally integrable, real-valued function on $[0, \infty)$ such that $\tilde{q}(x) \sim \frac{c}{x^2}$ $(x \to \infty)$ with constant c. Let*

$$c_{\operatorname{crit}} = \frac{a^2}{4|D|'(\lambda^{(n)})},$$

where D is the discriminant (1.5.6) of Hill's equation.

If $c/c_{\text{crit}} > 1$, then $\lambda^{(n)}$ is an accumulation point of eigenvalues in I_n of the perturbed periodic Sturm-Liouville operator

$$H_\alpha = -\frac{d}{dx}\left(p\frac{d}{dx}\right) + q(x) + \tilde{q}(x);$$

if $c/c_{\text{crit}} < 1$, then $\lambda^{(n)}$ is not an accumulation point of eigenvalues of H_α.

Proof. The perturbation matrix in the system (5.4.1) takes the form

$$\tilde{B} = \begin{pmatrix} -\tilde{q} & 0 \\ 0 & 0 \end{pmatrix} = \frac{1}{x^2}(-c + o(1))W,$$

and so $\hat{B} = -cW$. Hence the derivative of the discriminant $\tilde{D}$ of (5.4.14) can be expressed in terms of the derivative of D with respect to the spectral parameter,

$$|\tilde{D}|'(0) = -c|D|'(\lambda^{(n)}).$$

From Lemma 5.4.4 we see that

$$C_1 C_2 = \frac{c|D|'(\lambda^{(n)})}{a^2} = \frac{1}{4}\frac{c}{c_{\text{crit}}},$$

and so the function $\tilde{\phi}$ of Lemma 5.4.3 is bounded if $c/c_{\text{crit}} < 1$ and unbounded if $c/c_{\text{crit}} > 1$.

Now let u be a periodic or semi-periodic Floquet solution of the unperturbed system with spectral parameter $\lambda^{(n)}$ and w a solution of the perturbed system (5.4.1) with $\lambda = \lambda^{(n)}$, as in Lemma 5.4.1. Then, by Lemmas 5.4.1 and 5.4.2, the difference of the Prüfer angles of u and of w is globally bounded if $c/c_{\text{crit}} < 1$ and globally unbounded if $c/c_{\text{crit}} > 1$.

On the other hand, if z is a non-trivial solution of (5.4.1) with spectral parameter $\lambda \in I_n$, then by the reasoning in the proof of Theorem 5.3.1, its Prüfer angle differs by no more than a globally bounded error from that of a solution of the unperturbed equation with the same spectral parameter λ. The Prüfer angles of solutions of the unperturbed equation with spectral parameter in the closure $\overline{I_n}$ of the instability interval all have the same asymptotics (2.4.1), with only bounded errors.

Hence we conclude that the difference of the Prüfer angles of w and of z is globally bounded if $c/c_{\text{crit}} < 1$ and globally unbounded if $c/c_{\text{crit}} > 1$. The assertion of Theorem 5.4.5 now follows by the Relative Oscillation Theorem 4.8.3. □

We remark that, since the derivative of the discriminant has opposite sign at the two end-points of the same instability interval, at least one of the end-points is always in the subcritical case in the situation of Theorem 5.4.5. Hence eigenvalues can only accumulate at either the upper or the lower end of the gap in the essential spectrum, depending on the sign of c. By appeal to the comparison principle for Prüfer angles (Corollary 2.3.2), it is easy to see that perturbations which decay at

5.4. Critical coupling constants

a faster rate than x^{-2} only produce finitely many eigenvalues in any gap, whereas perturbations of fixed sign and slower decay rate than x^{-2} always generate an infinity of eigenvalues in each gap.

The behaviour in the borderline case $c = c_{\text{crit}}$ depends on higher-order asymptotics of the perturbation, as we explain in the Chapter notes.

For the perturbed periodic Dirac operator, the following analogous statement holds. As in section 5.3, the situation is complicated by the possibility of perturbing the matrix coefficients p_1 and p_2. We denote by $D(\lambda, c_1, c_2)$ the Hill discriminant of the periodic Dirac system (5.3.9). We emphasise that the critical constant in the next theorem plays a somewhat different role from that defined in Theorem 5.4.5.

Theorem 5.4.6. *Let $\lambda^{(n)}$ be an end-point of an instability interval I_n of the periodic Dirac equation (1.5.4), and let $\alpha \in [0, \pi)$. Let $\tilde{q}, \tilde{p}_1$ and $\tilde{p}_2$ be locally integrable, real-valued functions on $[0, \infty)$ such that*

$$\tilde{q}(x) \sim \frac{\hat{q}}{x^2}, \quad \tilde{p}_1(x) \sim \frac{\hat{p}_1}{x^2}, \quad \tilde{p}_2(x) \sim \frac{\hat{p}_2}{x^2} \quad (x \to \infty)$$

with constants $\hat{q}, \hat{p}_1, \hat{p}_2$. Let

$$c_{\text{crit}} = \frac{4}{a^2} \left(\hat{q} \frac{\partial}{\partial \lambda} |D|(\lambda^{(n)}, 0, 0) - \hat{p}_1 \frac{\partial}{\partial c_1} |D|(\lambda^{(n)}, 0, 0) - \hat{p}_2 \frac{\partial}{\partial c_2} |D|(\lambda^{(n)}, 0, 0) \right).$$

If $c_{\text{crit}} > 1$, then $\lambda^{(n)}$ is an accumulation point of eigenvalues in I_n of the perturbed periodic Dirac operator

$$H_\alpha = -i\sigma_2 \frac{d}{dx} + (p_1(x) + \tilde{p}_1(x)) \sigma_3 + (p_2(x) + \tilde{p}_2(x)) \sigma_1 + q(x) + \tilde{q}(x);$$

if $c_{\text{crit}} < 1$, then $\lambda^{(n)}$ is not an accumulation point of eigenvalues of H_α.

Proof. The matrix $\hat{B}$ of (5.4.5) takes the form

$$\hat{B} = \begin{pmatrix} -\hat{p}_1 - \hat{q} & -\hat{p}_2 \\ -\hat{p}_2 & \hat{p}_1 - \hat{q} \end{pmatrix}$$

and, comparing (5.4.14) and (5.3.9), we see that $\tilde{D}(c) = D(\lambda^{(n)} - c\hat{q}, c\hat{p}_1, c\hat{p}_2)$ and consequently

$$|\tilde{D}|'(0) = -\frac{\partial}{\partial \lambda}|D|(\lambda^{(n)}, 0, 0)\,\hat{q} + \frac{\partial}{\partial c_1}|D|(\lambda^{(n)}, 0, 0)\,\hat{p}_1 + \frac{\partial}{\partial c_2}|D|(\lambda^{(n)}, 0, 0)\,\hat{p}_2.$$

By Lemma 5.4.4, the function $\tilde{\phi}$ of Lemma 5.4.3 is globally bounded if $c_{\text{crit}} < 1$, and globally unbounded if $c_{\text{crit}} > 1$. By Lemmas 5.4.1 and 5.4.2, this also holds for the difference of the Prüfer angles of the solutions u of the unperturbed periodic equation and w of the perturbed periodic equation.

Inside the instability interval I_n, all the Prüfer angles of all solutions of the unperturbed equation have the same asymptotic (2.4.1) as u, with only bounded errors, and this also holds for the perturbed equation by the same comparison argument as in the proof of Theorem 5.3.2.

Thus the difference of the Prüfer angles of solutions of the perturbed periodic Dirac equation at $\lambda^{(n)}$ and at some $\lambda \in I_n$ is globally bounded if $c_{\text{crit}} < 1$ and globally unbounded if $c_{\text{crit}} > 1$, and the assertion of Theorem 5.4.6 follows by the Relative Oscillation Theorem 4.8.3. □

5.5 Eigenvalue asymptotics

We have seen in Theorem 5.4.5 for the perturbed Hill's equation and in Theorem 5.4.6 for the perturbed periodic Dirac equation that eigenvalues accumulate at an end-point of an instability interval of the unperturbed equation if the perturbation has x^{-2} decay with supercritical asymptotic constant. In the present section, we conclude the study of perturbed periodic problems by deriving the asymptotic distribution of the eigenvalues near their accumulation point. We shall use the oscillation techniques of section 5.4, but the reference equation will be the periodic equation with spectral parameter inside the instability interval, not at the end-point.

More precisely, let $\lambda^{(n)}$ be an end-point of a stability interval and λ' a fixed reference point inside the instability interval. We wish to count the eigenvalues between λ' and λ, where λ is a point between λ' and $\lambda^{(n)}$, and derive the leading asymptotics for this count in the limit $\lambda \to \lambda^{(n)}$. Since we assume the supercritical case, we know that the eigenvalue count will tend to infinity in the limit. By Theorems 5.3.1 or 5.3.2, it will be finite between any two points inside the same instability interval, and so the leading asymptotic is clearly independent of the choice of λ'.

Lemma 5.5.1. *Let $\lambda^{(n)}$ be an end-point of an instability interval I_n of the periodic equation (1.5.1). Then, for $\lambda \in I_n$, the Floquet exponent with positive real part satisfies*

$$\operatorname{Re}\mu(\lambda) = \sqrt{|D'(\lambda^{(n)})|}\sqrt{|\lambda - \lambda^{(n)}|} + o(\sqrt{|\lambda - \lambda^{(n)}|}) \qquad (\lambda \to \lambda^{(n)}), \qquad (5.5.1)$$

and the corresponding eigenvector $v(\lambda)$ of the monodromy matrix satisfies

$$v(\lambda) = v(\lambda^{(n)}) + O(\sqrt{|\lambda - \lambda^{(n)}|}) \qquad (\lambda \to \lambda^{(n)}). \qquad (5.5.2)$$

For the corresponding Floquet solution $u(x, \lambda)$, we have

$$u(x, \lambda) = u(x, \lambda^{(n)}) + O(\sqrt{|\lambda - \lambda^{(n)}|}) \qquad (\lambda \to \lambda^{(n)})$$

uniformly in $x \in [0, a]$, where $u(\cdot, \lambda^{(n)})$ is a-periodic or a-semi-periodic.

5.5. Eigenvalue asymptotics

Proof. The monodromy matrix $M(\lambda)$ is analytic in λ. In particular

$$M(\lambda) = M(\lambda^{(n)}) + M'(\lambda^{(n)})(\lambda - \lambda^{(n)}) + o(\lambda - \lambda^{(n)}), \qquad (5.5.3)$$

and similarly

$$D(\lambda) = D(\lambda^{(n)}) + D'(\lambda^{(n)})(\lambda - \lambda^{(n)}) + o(\lambda - \lambda^{(n)}) \qquad (5.5.4)$$

for the discriminant; here $|D(\lambda^{(n)})| = 2$. Since

$$|D(\lambda)| = 2\cosh\operatorname{Re}\mu(\lambda) \geq 2 + (\operatorname{Re}\mu(\lambda))^2$$

for $\lambda \in I_n$ by (1.4.6), we see from (5.5.4) that $(\operatorname{Re}\mu(\lambda))^2 = O(|\lambda - \lambda^{(n)}|)$, and hence

$$|D(\lambda)| = 2\cosh\operatorname{Re}\mu(\lambda) = 2 + (\operatorname{Re}\mu(\lambda))^2 + o(|\lambda - \lambda^{(n)}|),$$

which together with (5.5.4) yields (5.5.1).

Since we assume that $\lambda^{(n)}$ separates a stability interval from an instability interval and hence is not a point of coexistence, at least one of $\phi_{12}(\lambda), \phi_{21}(\lambda)$ in

$$M(\lambda) = \begin{pmatrix} \phi_{11}(\lambda) & \phi_{12}(\lambda) \\ \phi_{21}(\lambda) & \phi_{22}(\lambda) \end{pmatrix}$$

is non-zero at $\lambda^{(n)}$ and so, by continuity, also close to $\lambda^{(n)}$ in I_n. Assuming without loss of generality that $\phi_{21}(\lambda) \neq 0$, we have the eigenvector

$$v(\lambda) = \begin{pmatrix} e^{\mu(\lambda)} - \phi_{22}(\lambda) \\ \phi_{21}(\lambda) \end{pmatrix}$$

of $M(\lambda)$ for eigenvalue $e^{\mu(\lambda)}$. By (5.5.1),

$$e^{\mu(\lambda)} = \operatorname{sgn} D(\lambda^{(n)}) \, e^{\operatorname{Re}\mu(\lambda)} = \operatorname{sgn} D(\lambda^{(n)}) + O(\sqrt{|\lambda - \lambda^{(n)}|}).$$

Bearing in mind the asymptotics of $\phi_{21}(\lambda)$ and $\phi_{22}(\lambda)$ from (5.5.3), we conclude that (5.5.2) holds, where

$$v(\lambda^{(n)}) = \begin{pmatrix} \operatorname{sgn} D(\lambda^{(n)}) - \phi_{22}(\lambda^{(n)}) \\ \phi_{21}(\lambda^{(n)}) \end{pmatrix}$$

is an eigenvector of $M(\lambda^{(n)})$.

The statement about the Floquet solution now follows since, denoting by $\Phi(\cdot, \lambda)$ the canonical fundamental matrix of (1.5.1),

$$u(x, \lambda) = \Phi(x, \lambda) v(\lambda) = (\Phi(x, \lambda^{(n)}) + O(|\lambda - \lambda^{(n)}|))(v(\lambda^{(n)}) + O(\sqrt{|\lambda - \lambda^{(n)}|})$$
$$= u(x, \lambda^{(n)}) + O(\sqrt{|\lambda - \lambda^{(n)}|})$$

uniformly in $x \in [0, a]$. $\square$

We now study the perturbed equation (5.4.1) on $[0, \infty)$, where the perturbation takes the form (5.4.5) for $x \geq 1$. For all λ, let $\theta(x, \lambda)$ be the Prüfer angle of a real-valued solution, with $\theta(0, \lambda) = \alpha$; here $\alpha \in [0, \pi)$ is the parameter of the boundary condition at 0. By the Relative Oscillation Theorem 4.8.3, we only need to estimate $\theta(x, \lambda) - \theta(x, \lambda^{(n)})$ in the limit as $x \to \infty$ in order to count the eigenvalues between λ and λ' up to a bounded error.

Since the perturbation $\tilde{B}(x)$ tends to 0 as $x \to \infty$, it follows by Sturm Comparison, as in the proof of Theorem 5.3.2, that $\theta(x, \lambda)$ grows regularly as $\frac{n\pi x}{a}$, up to a bounded error, as soon as x is so large that $|\tilde{B}(x)| \leq |\lambda - \lambda^{(n)}|$. Therefore we need not keep track of $\theta(x, \lambda)$ for all $x \geq 0$, but only up to a λ-dependent point $r(\lambda)$, after which the difference $\theta(x, \lambda) - \theta(x, \lambda')$ will be bounded uniformly in λ. The leading asymptotic of the number of eigenvalues between λ and λ' will be determined by the growth of $\theta(r(\lambda), \lambda) - \theta(r(\lambda), \lambda')$ as $\lambda \to \lambda^{(n)}$.

Note that we can assume without loss of generality that the matrix $\hat{B}$ determining the asymptotic behaviour of the perturbation $\tilde{B}$ is non-zero in the following. Indeed, $\hat{B} = 0$ would mean that $F_2 = 0$ in (5.4.8) and hence $C_2 = 0$ in Lemma 5.4.2. Then $C_1 C_2 = 0 < 1/4$ in Lemma 5.4.3; this is the subcritical case without accumulation of eigenvalues at $\lambda^{(n)}$ and not of interest here.

Lemma 5.5.2. *Let $\hat{B}$ be a non-zero symmetric 2×2 matrix with real entries and β a symmetric 2×2 matrix-valued function on $[0, \infty)$ with $\lim_{x \to \infty} \beta(x) = 0$.*

Then for $\lambda \in I_n$ there is $r(\lambda) \geq 1$ such that

$$\left| \frac{1}{x^2} (\hat{B}(x) + \beta(x)) \right| \leq |\lambda - \lambda^{(n)}| \qquad (x \geq r(\lambda))$$

and

$$r(\lambda) \sim \frac{\sqrt{|\hat{B}|}}{\sqrt{|\lambda - \lambda^{(n)}|}} \qquad (\lambda \to \lambda^{(n)}). \tag{5.5.5}$$

Proof. First let

$$r_1(\lambda) := \frac{\sqrt{|\hat{B}|}}{\sqrt{|\lambda - \lambda^{(n)}|}} + 1,$$

then set

$$r(\lambda) := \frac{\sqrt{|\hat{B}| + \sup_{x \geq r_1(\lambda)} |\beta(x)|}}{\sqrt{|\lambda - \lambda^{(n)}|}} + 1 \geq r_1(\lambda).$$

Then for $x \geq r(\lambda)$,

$$\left| \frac{1}{x^2}(\hat{B} + \beta(x)) \right| \leq \frac{|\hat{B}| + |\beta(x)|}{|\hat{B}| + \sup_{x \geq r_1(\lambda)} |\beta(x)|} |\lambda - \lambda^{(n)}| \leq |\lambda - \lambda^{(n)}|,$$

5.5. Eigenvalue asymptotics

as required. Moreover,

$$0 \leq r(\lambda) - r_1(\lambda) + 1 \leq \frac{\sup\limits_{x \geq r_1(\lambda)} |\beta(x)|}{\sqrt{|\lambda - \lambda^{(n)}|} \, 2\sqrt{|\hat{B}| + 1}} + 1 = o\left(\frac{1}{\sqrt{|\lambda - \lambda^{(n)}|}}\right)$$

$(\lambda \to \lambda^{(n)})$, which proves (5.5.5). $\square$

Clearly, $r(\lambda)$ as given in the above lemma tends to ∞ as $\lambda \to \lambda^{(n)}$. Taking into account the regular growth behaviour of $\theta(r(\lambda), \lambda')$, we conclude that, up to an error bounded uniformly in $\lambda \in I_n$, the number of eigenvalues between λ and λ' is given by

$$\frac{1}{\pi} \theta(r(\lambda), \lambda) - \frac{nr(\lambda)}{a}. \tag{5.5.6}$$

On the other hand, we know from Theorem 2.4.1 that the Prüfer angle $\theta_0(x, \lambda)$ of a real-valued solution of the unperturbed periodic equation (1.5.1), with $\theta_0(0, \lambda) \in [0, \pi)$, also satisfies

$$\theta_0(x, \lambda) = \frac{n\pi x}{a} + O_{\text{unif}}(1),$$

with error term bounded uniformly in $\lambda \in \overline{I_n}$. Thus the difference of angles in (5.5.6) can be read as the relative rotation, up to the point $r(\lambda)$, of a solution of the perturbed equation with spectral parameter λ compared to a solution of the unperturbed equation with the same spectral parameter. By Lemma 5.4.1, this difference can be estimated, up to a universally bounded error, by the growth of a solution γ of (5.4.3).

We now follow the general approach of section 5.4 from (5.4.5) onwards, but with the difference that the solution u of the unperturbed equation will not be periodic or semi-periodic, but a Floquet solution with Floquet multiplier of modulus $|e^{\mu(\lambda)}| > 1$.

For the coefficients f and g in (5.4.6) — which now depend on λ, too —, we have

$$f(x) = -|u(x, \lambda)|^{-2}, \qquad g'(x) = |u(x, \lambda)|^{-4} \, u(x, \lambda)^T (JA(x, \lambda) - A(x, \lambda)J) u(x, \lambda)$$

from (1.9.8) and (1.9.9); here $A(x, \lambda) = J(B(x) + \lambda W(x))$. As $u(x, \lambda)$ is a Floquet solution with multiplier $e^{\mu(\lambda)}$ and $\lambda \in I_n$, (1.4.7) shows that $e^{-\operatorname{Re}\mu(\lambda)x/a} u(x, \lambda)$ is a-periodic or a-semi-periodic. Therefore the functions $f(x) \, e^{2\operatorname{Re}\mu(\lambda)x/a}$ and $g'(x) \, e^{2\operatorname{Re}\mu(\lambda)x/a}$ are a-periodic.

In the light of this observation, we now start from (5.4.7), perform the Kepler transformation

$$\tan \psi(x) = \frac{e^{2\operatorname{Re}\mu(\lambda)x/a}}{x} \left(\tan \gamma(x) + g(x)\right)$$

and obtain the differential equation for ψ,

$$\psi' = \frac{2\operatorname{Re}\mu(\lambda)}{a} \sin\psi\cos\psi$$
$$+ \frac{1}{x}\left(e^{2\operatorname{Re}\mu(\lambda)x/a} g'(x)\cos^2\psi - \sin\psi\cos\psi \right.$$
$$\left. + e^{-2\operatorname{Re}\mu(\lambda)x/a} u^T(\hat{B}+\beta(x))u\sin^2\psi \right)$$
$$+ \frac{1}{x^2} f(x)\left((Ju)^T(\hat{B}+\beta(x))u + u^T(\hat{B}+\beta(x))Ju\right)\sin\psi\cos\psi$$
$$+ \frac{1}{x^3} e^{2\operatorname{Re}\mu(\lambda)x/a} f^2(x)(Ju)^T(\hat{B}+\beta(x))Ju\cos^2\psi. \qquad (5.5.7)$$

Defining the a-periodic functions

$$F_1(x,\lambda) = g'(x)\,e^{2\operatorname{Re}\mu(\lambda)x/a}, \qquad F_2(x,\lambda) = e^{-2\operatorname{Re}\mu(\lambda)x/a} u^T(x,\lambda)\,\hat{B}\,u(x,\lambda)$$

we can rewrite (5.5.7) in the form

$$\psi' = \frac{\operatorname{Re}\mu(\lambda)}{a}\sin 2\psi + \frac{1}{x}\left(F_1(x,\lambda)\cos^2\psi - \sin\psi\cos\psi + F_2(x,\lambda)\sin^2\psi\right)$$
$$+ \frac{e^{-2\operatorname{Re}\mu(\lambda)x/a}}{x} u^T(\hat{B}+\beta(x))u\sin^2\psi + O_{\text{unif}}\left(\frac{1}{x^2}\right). \qquad (5.5.8)$$

The remainder term is uniform in $\lambda \in \overline{I_n}$; indeed, the combinations of f, u and the exponential in the last two terms of (5.5.7) are a-periodic, hence bounded, in x and continuous in λ.

We shall now apply an averaging procedure analogous to that of Lemma 5.4.2. There is the essential difference that (5.5.8) has a non-decaying leading term; however, we are here concerned with the limit $\lambda \to \lambda^{(n)}$, and the x decay enters only indirectly as $r(\lambda) \to \infty$ in that limit.

We now define

$$\tilde{\psi}(x) = \frac{1}{a}\int_x^{x+a} \psi \qquad (x \geq 1)$$

and apply the Mean Value Theorem for integrals, which gives, for each $x \geq 1$, an $x' \in [x, x+a]$ such that $\tilde{\psi}(x) = \psi(x')$. We then obtain

$$|\psi(t) - \tilde{\psi}(x)| = \left|\int_{x'}^t \psi'\right|$$
$$\leq \operatorname{Re}\mu(\lambda) + \frac{1}{x}\int_0^a \left(|F_1(t,\lambda)| + \frac{1}{2} + |F_2(t,\lambda)|\right)dt + o_{\text{unif}}\left(\frac{1}{x}\right)$$
$$= \operatorname{Re}\mu(\lambda) + O_{\text{unif}}\left(\frac{1}{x}\right) \qquad (5.5.9)$$

5.5. Eigenvalue asymptotics

for all $t \in [x, x+a]$. By (5.5.1) and (5.5.5), this implies that

$$|\psi(r(\lambda), \lambda) - \tilde{\psi}(r(\lambda), \lambda)| \leq \operatorname{Re} \mu(\lambda) + O_{\text{unif}}\left(\frac{1}{r(\lambda)}\right) = O(\sqrt{|\lambda - \lambda^{(n)}|}) \quad (\lambda \to \lambda^{(n)}).$$

Also, $|\psi(0, \lambda) - \tilde{\psi}(0, \lambda)| = O(1)$, and so $\tilde{\psi}$ can be used instead of ψ for the purpose of counting eigenvalues up to bounded error. Defining the continuous functions

$$C_j(\lambda) = \frac{1}{a}\int_0^a F_j(t, \lambda)\, dt \quad (\lambda \in \overline{I_n}; j \in \{1, 2\}),$$

we find from (5.5.8) that

$$\tilde{\psi}'(x) = \frac{1}{a}\int_x^{x+a} \psi'$$

$$= \frac{\operatorname{Re}\mu(\lambda)}{a^2}\int_x^{x+a}\sin 2\psi + \frac{1}{x}\left(C_1(\lambda)\cos^2\tilde{\psi} - \sin\tilde{\psi}\cos\tilde{\psi} + C_2(\lambda)\sin^2\tilde{\psi}\right)$$

$$+ \frac{1}{xa}\int_x^{x+a} e^{-2\operatorname{Re}\mu(\lambda)t/a}\, u(t,\lambda)^T\beta(t)u(t,\lambda)\sin^2\psi(t,\lambda)\, dt$$

$$+ \frac{1}{xa}\int_x^{x+a}\Big(F_1(t,\lambda)\left(\cos^2\psi - \cos^2\tilde{\psi}\right)$$

$$- (\sin\psi\cos\psi - \sin\tilde{\psi}\cos\tilde{\psi}) + F_2(t,\lambda)\left(\sin^2\psi - \sin^2\tilde{\psi}\right)\Big)\, dt$$

$$+ O_{\text{unif}}\left(\frac{1}{x^2}\right). \tag{5.5.10}$$

In view of (5.5.1), the first term on the right-hand side of (5.5.10) is of order $O(\sqrt{|\lambda - \lambda^{(n)}|})$. For the last integral in (5.5.10), we use (5.4.11) and (5.5.9) together with the boundedness, uniformly in $\lambda \in I_n$, of F_1 and F_2 to obtain the estimate

$$\operatorname{Re}\mu(\lambda)\, O_{\text{unif}}\left(\frac{1}{x}\right) + O_{\text{unif}}\left(\frac{1}{x^2}\right) = O(\sqrt{|\lambda - \lambda^{(n)}|}) + O_{\text{unif}}\left(\frac{1}{x^2}\right).$$

Similarly, $e^{-2\operatorname{Re}\mu(\lambda)x/a}\, u^T\beta u = o_{\text{unif}}(1)$. Thus we can write (5.5.10) more briefly in the form

$$\tilde{\psi}'(x) = \frac{1}{x}\left(C_1(\lambda)\cos^2\tilde{\psi} - \sin\tilde{\psi}\cos\tilde{\psi} + C_2(\lambda)\sin^2\tilde{\psi}\right) + O(\sqrt{|\lambda - \lambda^{(n)}|}) + o_{\text{unif}}\left(\frac{1}{x}\right).$$

The asymptotics of Lemma 5.5.1 for $u(x, \lambda)$ and $\mu(\lambda)$, along with

$$A(x, \lambda) = A(x, \lambda^{(n)}) + (\lambda - \lambda^{(n)})JW,$$

give

$$C_j(\lambda) = C_j(\lambda^{(n)}) + O(\sqrt{|\lambda - \lambda^{(n)}|}) \quad (\lambda \to \lambda^{(n)}; j \in \{1, 2\}). \tag{5.5.11}$$

We can now deduce the asymptotics of $\tilde{\phi}$ from the following lemma.

Lemma 5.5.3. *Let Λ be a closed interval with end-point $\lambda^{(n)}$, and assume that $C_1, C_2 : \Lambda \to \mathbb{R}$ are continuous and satisfy (5.5.11) and*

$$4\, C_1(\lambda^{(n)})\, C_2(\lambda^{(n)}) > 1. \tag{5.5.12}$$

Moreover, let $\tilde{\psi} : [1, \infty) \times \Lambda \to \mathbb{R}$ be a function such that $\tilde{\psi}(\cdot, \lambda)$ is locally absolutely continuous for each $\lambda \in \Lambda$ and

$$\tilde{\psi}'(x, \lambda) = \frac{1}{x}\left(C_1(\lambda) \cos^2 \tilde{\psi}(x, \lambda) - \sin \tilde{\psi}(x, \lambda) \cos \tilde{\psi}(x, \lambda) + C_2(\lambda) \sin^2 \tilde{\psi}(x, \lambda)\right)$$
$$+ G(x, \lambda) \tag{5.5.13}$$

with

$$|G(x, \lambda)| \le c\sqrt{|\lambda - \lambda^{(n)}|} + h(x) \qquad (x \ge 1, \lambda \in \Lambda), \tag{5.5.14}$$

where $c > 0$ is a constant, $h > 0$ and $\lim_{x \to \infty} x\, h(x) = 0$. Also assume that $r(\lambda)$ has asymptotic growth (5.5.5). Then

$$|\tilde{\psi}(r(\lambda), \lambda) - \tilde{\psi}(1, \lambda)| \sim \frac{1}{4}\sqrt{4\, C_1(\lambda^{(n)})\, C_2(\lambda^{(n)}) - 1}\, \left|\log|\lambda - \lambda^{(n)}|\right| \qquad (\lambda \to \lambda^{(n)}).$$

Proof. In analogy to the beginning of the proof of Lemma 5.4.3, we choose a continuous $\psi_0 : \Lambda \to \mathbb{R}$ such that

$$\sin 2\psi_0(\lambda) = \frac{-1}{\sqrt{1 + (C_1(\lambda) - C_2(\lambda))^2}}, \quad \cos 2\psi_0(\lambda) = \frac{C_1(\lambda) - C_2(\lambda)}{\sqrt{1 + (C_1(\lambda) - C_2(\lambda))^2}},$$

and rewrite (5.5.13) as

$$\omega'(x, \lambda) = \frac{\Gamma_+(\lambda)}{2x}\left(\cos^2 \omega(x, \lambda) + \frac{\Gamma_-(\lambda)}{\Gamma_+(\lambda)} \sin^2 \omega(x, \lambda)\right) + G(x, \lambda),$$

where $\omega(x, \lambda) = \tilde{\psi}(x, \lambda) - \psi_0(\lambda)$, ω' denotes the derivative with respect to x, and

$$\Gamma_\pm(\lambda) = C_1(\lambda) + C_2(\lambda) \pm \sqrt{1 + (C_1(\lambda) - C_2(\lambda))^2}.$$

By (5.5.12) and continuity, $|C_1(\lambda) + C_2(\lambda)| > \sqrt{1 + (C_1(\lambda) - C_2(\lambda))^2}$ and hence

$$\frac{\Gamma_-(\lambda)}{\Gamma_+(\lambda)} > 0$$

for λ sufficiently close to $\lambda^{(n)}$, and for such λ we can perform the Kepler transformation

$$\tan \tilde{\omega} = \arctan\left(\sqrt{\frac{\Gamma_-(\lambda)}{\Gamma_+(\lambda)}}\, \tan \omega\right).$$

5.5. Eigenvalue asymptotics

Then, by Theorem 2.2.1,

$$\tilde{\omega}'(x,\lambda) = \sqrt{\frac{\Gamma_-(\lambda)}{\Gamma_+(\lambda)}} \left(\frac{\Gamma_+(\lambda)}{2x} + \frac{G(x,\lambda)}{\cos^2 \omega(x,\lambda) + \frac{\Gamma_-(\lambda)}{\Gamma_+(\lambda)} \sin^2 \omega(x,\lambda)} \right)$$

$$= \frac{\operatorname{sgn}(C_1(\lambda) + C_2(\lambda))}{2x} \sqrt{4\,C_1(\lambda)\,C_2(\lambda) - 1}$$

$$+ \left(\sqrt{\frac{\Gamma_-(\lambda)}{\Gamma_+(\lambda)}} \cos^2 \tilde{\omega}(x,\lambda) + \sqrt{\frac{\Gamma_+(\lambda)}{\Gamma_-(\lambda)}} \sin^2 \tilde{\omega}(x,\lambda) \right) G(x,\lambda).$$
(5.5.15)

By (5.5.14), we can estimate

$$\left| \int_1^{r(\lambda)} \left(\sqrt{\frac{\Gamma_-(\lambda)}{\Gamma_+(\lambda)}} \cos^2 \tilde{\omega}(x,\lambda) + \sqrt{\frac{\Gamma_+(\lambda)}{\Gamma_-(\lambda)}} \sin^2 \tilde{\omega}(x,\lambda) \right) G(x,\lambda)\,dx \right|$$

$$\leq S(\lambda)\, c\sqrt{|\lambda - \lambda^{(n)}|}\, \frac{\sqrt{|\hat{B}| + \sup_{x \geq r_1(\lambda)} |\beta(x)|}}{\sqrt{|\lambda - \lambda^{(n)}|}} + S(\lambda) \int_1^{r(\lambda)} h,$$

where $S = \sqrt{\frac{\Gamma_-}{\Gamma_+}} + \sqrt{\frac{\Gamma_+}{\Gamma_-}}$. Both $S(\lambda)$ and $|\hat{B}| + \sup_{x \geq r_1(\lambda)} |\beta(x)|$ remain bounded as $\lambda \to \lambda^{(n)}$, and by l'Hospital's rule

$$\lim_{x \to \infty} \frac{\int_1^x h}{\log x} = \lim_{x \to \infty} x\,h(x) = 0,$$

and so

$$\int_1^{r(\lambda)} h = o(\log r(\lambda)) \qquad (\lambda \to \lambda^{(n)}).$$

From (5.5.11),

$$\sqrt{4\,C_1(\lambda)\,C_2(\lambda) - 1} = \sqrt{4\,C_1(\lambda^{(n)})\,C_2(\lambda^{(n)}) - 1} + O(\sqrt{|\lambda - \lambda^{(n)}|}) \qquad (\lambda \to \lambda^{(n)}),$$

and

$$\log r(\lambda) = -\frac{1}{2} \log |\lambda - \lambda^{(n)}| + \log \left(\sqrt{|\hat{B}| + \sup_{x \geq r_1(\lambda)} |\beta(x)|} + \sqrt{|\lambda - \lambda^{(n)}|} \right)$$

$$= -\frac{1}{2} \log |\lambda - \lambda^{(n)}| + O(1) \qquad (\lambda \to \lambda^{(n)}).$$

Thus, integrating (5.5.15) we obtain the asymptotic

$$\tilde{\omega}(r(\lambda),\lambda) - \tilde{\omega}(1,\lambda)$$
$$= \frac{\operatorname{sgn}(C_1(\lambda) + C_2(\lambda))}{2} \sqrt{4\,C_1(\lambda)\,C_2(\lambda) - 1}\,\log r(\lambda) + o(\log r(\lambda))$$
$$\sim -\frac{\operatorname{sgn}(C_1(\lambda^{(n)}) + C_2(\lambda^{(n)}))}{4} \sqrt{4\,C_1(\lambda^{(n)})\,C_2(\lambda^{(n)}) - 1}\,\log|\lambda - \lambda^{(n)}|.$$

$\square$

We have thus proved the following statements about perturbed periodic Sturm-Liouville and Dirac operators in the supercritical case.

Theorem 5.5.4. *Let $\lambda^{(n)}$ be an end-point of an instability interval I_n of Hill's equation (1.5.2) with $w = 1$, and let $\lambda' \in I_n$. Moreover, let $\alpha \in [0,\pi)$ and let $\tilde{q}$ be a locally integrable, real-valued function on $[0,\infty)$ of asymptotic $\tilde{q}(x) \sim \frac{c}{x^2}$ ($x \to \infty$) with constant c such that $c/c_{\text{crit}} > 1$, where*

$$c_{\text{crit}} = \frac{a^2}{4|D|'(\lambda^{(n)})},$$

and D is the discriminant (1.5.6) of Hill's equation.

Then the number $N(\lambda)$ of eigenvalues between λ' and λ of the perturbed periodic Sturm-Liouville operator

$$H_\alpha = -\frac{d}{dx}\left(p\frac{d}{dx}\right) + q(x) + \tilde{q}(x)$$

has asymptotic

$$N(\lambda) \sim \frac{1}{4\pi}\sqrt{\frac{c}{c_{\text{crit}}} - 1}\,\Big|\log|\lambda - \lambda^{(n)}|\Big| \qquad (\lambda \to \lambda^{(n)}). \qquad (5.5.16)$$

Theorem 5.5.5. *Let $\lambda^{(n)}$ be an end-point of an instability interval I_n of the periodic Dirac equation (1.5.4), and let $\lambda' \in I_n$ and $\alpha \in [0,\pi)$. Let $\tilde{q}, \tilde{p}_1$ and $\tilde{p}_2$ be locally integrable, real-valued functions on $[0,\infty)$ of asymptotic*

$$\tilde{q}(x) \sim \frac{\hat{q}}{x^2}, \quad \tilde{p}_1(x) \sim \frac{\hat{p}_1}{x^2}, \quad \tilde{p}_2(x) \sim \frac{\hat{p}_2}{x^2} \qquad (x \to \infty)$$

with constants $\hat{q}, \hat{p}_1, \hat{p}_2$ such that

$$c_{\text{crit}} = \frac{4}{a^2}\left(\hat{q}\frac{\partial}{\partial\lambda}|D|(\lambda^{(n)},0,0) - \hat{p}_1\frac{\partial}{\partial c_1}|D|(\lambda^{(n)},0,0) - \hat{p}_2\frac{\partial}{\partial c_2}|D|(\lambda^{(n)},0,0)\right) > 1.$$

Then the number $N(\lambda)$ of eigenvalues between λ' and λ of the perturbed periodic Dirac operator

$$H_\alpha = -i\sigma_2\frac{d}{dx} + (p_1(x) + \tilde{p}_1(x))\,\sigma_3 + (p_2(x) + \tilde{p}_2(x))\,\sigma_1 + q(x) + \tilde{q}(x);$$

has asymptotic

$$N(\lambda) \sim \frac{1}{4\pi} \sqrt{c_{\text{crit}} - 1} \left| \log|\lambda - \lambda^{(n)}| \right| \qquad (\lambda \to \lambda^{(n)}).$$

5.6 Chapter notes

§5.2 Corollary 5.2.3 was shown by Stolz [175] for the one-dimensional Schrödinger operator, i.e. $w = p = 1$, $\tilde{p} = 0$; the proof of Theorem 5.2.1 follows his idea. It is worth noting Stolz's remark that the conditions (5.2.1) and (5.2.2) — and the ensuing conditions on the perturbations of the coefficients of the Dirac and Sturm-Liouville operators — are satisfied by absolutely integrable functions, functions of bounded variation multiplied with a-periodic functions, and linear combinations of these. The results of this section carry over to the case where the left-hand end-point is singular, in particular to the full-line operator, see [175].

§5.3 The results of this section extend analogously to the situation with two singular end-points by Glazman's decomposition method [67, Section 7]. If the Sturm-Liouville or Dirac operator is given on an interval (a, b) with both a and b singular, we can choose a point $c \in (a, b)$ and consider the two boundary-value problems on $(a, c]$ and on $[c, \infty)$ with some boundary conditions at c. The operator on (a, b) is then a two-dimensional extension of the direct sum of one-dimensional restrictions of the two part-interval operators with the boundary condition strengthened to the condition that $y(c) = (py')(c) = 0$ for Sturm-Liouville and $u(c) = 0$ for Dirac. Using the spectral representation, one can hence deduce that the total multiplicity of the spectrum of the operator on (a, b) differs by no more than 4 from the sum of the spectral multiplicities of the part-interval operators.

Theorem 5.3.3 is due to Sobolev [170], who considers a strong-coupling limit of a perturbation of inverse power decay, which is clearly equivalent to the slow-variation limit. Theorem 5.3.4 can be found in [164] for the specific case of perturbations of the type of the angular momentum term; there the perturbation is singular at 0 but can be shown to generate no eigenvalues in any given compact λ-interval when considered on $(0, c]$ with sufficiently small c.

Numerical evidence suggests that, in the case of the angular momentum perturbation, the asymptotic densities (5.3.5) and (5.3.10) can give a very accurate indication of the distribution of eigenvalues in a gap even for small scaling constants c [25], [163]. The growing density of eigenvalues in each gap with increasing angular momentum quantum number (corresponding to increasing c) explains the appearance of dense point spectrum in radially periodic higher-dimensional Schrödinger and Dirac operators [83], [159].

A more precise count of eigenvalues in distant gaps of Hill's equation under the assumption that the perturbation satisfies $\int_\mathbb{R} (1 + |x|) \tilde{q}(x)\, dx < \infty$ was announced by Rofe-Beketov [150]: all sufficiently distant gaps contain at most 2 eigenvalues, at least one eigenvalue if $\int_\mathbb{R} \tilde{q} \neq 0$, and exactly one eigenvalue if in addition $\tilde{q} \geq 0$. See [63] for further details; a refinement of these results was given in [21]. The existence of eigenvalues in spectral gaps of perturbed Schrödinger operators was also shown by Deift and Hempel [32] in a more general setting; see also [62].

§**5.4** Theorem 5.4.5 goes back to Rofe-Beketov (first announced in [151, Theorems 9, 10], [152, Theorems 4, 5]; details in [154], [153] including extensions to almost periodic equations); he treated the differential equation (5.4.3) in the Sturm-Liouville case by relating it to a different Sturm-Liouville equation by means of an implicit variable transformation. The oscillation properties of the resulting equation can then be studied using a criterion of Taam, Hille and Wintner [122, Theorem B]. The approach presented here [161] can also be applied to decide the question of finiteness or infinity of the gap spectrum in the limiting case $c = c_{\text{crit}}$ of Theorem 5.4.5. If $\tilde{q}(x) \sim \frac{c_{\text{crit}}}{x^2} + \frac{c}{x^2(\log x)^2}$, then $\lambda^{(n)}$ is an accumulation point of eigenvalues if $c/c_{\text{crit}} > 1$ and not an accumulation point of eigenvalues if $c/c_{\text{crit}} < 1$ [161, Proposition 4] (in particular, the case $\tilde{q}(x) = \frac{c_{\text{crit}}}{x^2}$ is subcritical). This extends to a whole hierarchy of asymptotic terms with associated critical coupling constants, see [121], analogous to the classical oscillation criterion of Kneser [111] as refined by Weber, Hartman and Hille (cf. [79, Chapter XI, Exercise 1.2]) — this is the special case of Theorem 5.4.5 where the periodic coefficients are constant and $\lambda^{(n)}$ is the infimum of the essential spectrum (in particular, $n = 0$). If $p = w = 1$ and q is constant, then $c_{\text{crit}} = -1/4$.

This corresponds to the constant in Hardy's inequality. Indeed, separation of variables in polar coordinates of the Laplacian $-\Delta$ in $\mathbb{R}^2$ gives rise to a direct sum decomposition of this operator with half-line Sturm-Liouville operators

$$-\frac{d^2}{dr^2} + \frac{\ell^2 - \frac{1}{4}}{r^2}$$

as terms, where $\ell \in \{0, 1, 2, \dots\}$ and r is the radial variable; and $\ell = 0$ is just the borderline case of Theorem 5.4.5 — which must be subcritical since the Laplacian has no eigenvalues. A curious phenomenon appears when a non-constant radially periodic potential $q(r)$ is added to the Laplacian [160]: if λ_0 is the infimum of the essential spectrum, the Floquet solution y corresponding to $u = \begin{pmatrix} y \\ y' \end{pmatrix}$ can be chosen strictly positive, and d'Alembert's formula (1.9.3) can be used instead of the Rofe-Beketov formula (5.4.6), giving $F_1 = 1/y^2$. Then

$$C_1 C_2 = \frac{-c}{a^2} \int_0^a y^2 \int_0^a \frac{1}{y^2},$$

and it follows by the Cauchy-Schwarz inequality that $c_{\text{crit}} \geq -\frac{1}{4}$ with equality if and only if y^2 and y^{-2} are linearly dependent, i.e. if q is constant. Hence the two-dimensional Schrödinger operator with non-constant rotationally symmetric potential always has infinitely many eigenvalues below the essential spectrum.

We note that results similar to Theorem 5.4.5 in a slightly more general setting can be found in [109]. Gesztesy and Ünal [65] extended Kneser's oscillation

5.6. Chapter notes

criterion and its relation to Hardy's inequality to allow background potentials, including the periodic case.

The relative oscillation result underlying Theorem 5.4.6 appears in [162], see also [172]. The critical case also gives rise to a ladder of asymptotic scales for the perturbations, as for the Sturm-Liouville equation.

§**5.5** Theorem 5.5.4 was proven in [161] by the method shown in section 5.5. The analogous statement for the perturbed periodic Dirac operator, Theorem 5.5.5, is new.

It is of interest to compare the asymptotics of eigenvalues in instability intervals with the semiclassical asymptotics. In the absence of a periodic background potential in the Sturm-Liouville equation, i.e. if $q = 0$ in $H_\alpha = -\frac{d^2}{dx^2} + q(x) + \tilde{q}(x)$, the semiclassical formula for the number of eigenvalues below λ,

$$N(\lambda) \sim \frac{1}{\pi} \int_0^\infty \sqrt{(\lambda - \tilde{q}(x))_+}\, dx$$

(where $(\lambda - \tilde{q}(x))_+ = \max\{0, \lambda - q(x)\}$) is known to hold true in a variety of situations ([135], [185, Chapter VII], [78], [124], [157]). However, it is clear that this formula cannot be true in the case $\tilde{q}(x) \sim c/x^2$ ($x \to \infty$), because the finiteness or otherwise of the limit of the above semiclassical integral as $\lambda \to 0$ is independent of c, but there is a critical constant $c_{\text{crit}} = 1/4$ in the sense of Theorem 5.4.5. For this case, Jörgens [104] has given the adjusted asymptotics

$$N(\lambda) \sim \frac{1}{\pi} \int_1^\infty \sqrt{\left(\lambda - \tilde{q}(x) - \frac{1}{4x^2}\right)_+}\, dx \qquad (\lambda \to 0).$$

Note that here the integrand for $\lambda < 0$ has compact support in the subcritical case.

For the perturbed periodic Sturm-Liouville operator with non-constant periodic potential of period 1 and perturbation $\tilde{q} \asymp -x^{-b}$, $0 < b < 1/n$, Zelenko derived an m-term asymptotic series for the number of eigenvalues close to an end-point $\lambda^{(n)}$ of an instability interval [208, Thm. 2]; for $m = 1$ it has the form

$$N(\lambda) \sim \frac{\sqrt{|D'(\lambda^{(n)})|}}{\pi} \int_0^\infty \sqrt{(\lambda - \lambda^{(n)} - \tilde{q}(x))_+}\, dx \qquad (\lambda \to \lambda^{(n)}).$$

For the same reason as above, this formula must fail when $b = 0$. For comparison, we note that for $\tilde{q}(x) = c/x^2 + O(1/x^{2+\epsilon})$ with $\epsilon > 0$ and supercritical c, (5.5.16) can be rewritten as

$$N(\lambda) \sim \frac{\sqrt{|D'(\lambda^{(n)})|}}{\pi a} \int_1^\infty \sqrt{\left(\left|\tilde{q}(x) - \frac{c_{\text{crit}}}{x^2}\right| - |\lambda - \lambda^{(n)}|\right)_+}\, dx \qquad (\lambda \to \lambda^{(n)}).$$

Thus this asymptotic formula closes the gap between Jörgens' and Zelenko's asymptotics, generalising the former by the inclusion of a periodic background, the latter by allowing critical decay of the perturbation.

Bibliography

[1] N. I. Akhiezer, *A continuous analogue of orthogonal polynomials on a system of intervals*, Soviet Math **2** (1961), 1409–1412.

[2] S. I. Al′ber, *Investigation of equations of Korteweg-de Vries type by the method of recurrence relations*, J. London Math. Soc. (2) **19** (1979), no. 3, 467–480. MR 540063 (81b:35084)

[3] J. M. Almira and P. J. Torres, *Invariance of the stability of Meissner's equation under a permutation of its intervals*, Ann. Mat. Pura Appl. (4) **180** (2001), no. 2, 245–253. MR 1847407 (2002e:34054)

[4] B. Anahtarci and P. Djakov, *Improved asymptotics of the spectral gap for the Mathieu operator*, Preprint arXiv:1202.4623 (2012).

[5] F. M. Arscott, *Periodic differential equations. An introduction to Mathieu, Lamé, and allied functions*, International Series of Monographs in Pure and Applied Mathematics, Vol. 66. A Pergamon Press Book, The Macmillan Co., New York, 1964. MR 0173798 (30 #4006)

[6] F. V. Atkinson, *Estimation of an eigenvalue occurring in a stability problem*, Math. Z. **68** (1957), 82–99. MR 0092044 (19,1052a)

[7] _____, *Discrete and continuous boundary problems*, Mathematics in Science and Engineering, Vol. 8, Academic Press, New York, 1964. MR 0176141 (31 #416)

[8] J. Avron and B. Simon, *The asymptotics of the gap in the Mathieu equation*, Ann. Physics **134** (1981), no. 1, 76–84. MR 626698 (82h:34030)

[9] A. Badanin, J. Brüning, and E. Korotyaev, *The Lyapunov function for Schrödinger operators with a periodic 2×2 matrix potential*, J. Funct. Anal. **234** (2006), no. 1, 106–126. MR 2214141 (2006k:47090)

[10] A. Badanin and E. Korotyaev, *Spectral asymptotics for periodic fourth-order operators*, Int. Math. Res. Not. (2005), no. 45, 2775–2814. MR 2182471 (2006f:34064)

[11] M. Bell, *A note on Mathieu functions*, Proc. Glasgow Math. Assoc. **3** (1957), 132–134. MR 0120408 (22 #11162)

[12] E. D. Belokolos, A. I. Bobenko, V. Z. Enol'skii, and A. R. Its, *Algebro-geometric approach to nonlinear integrable equations (Springer series in nonlinear dynamics)*, Springer, Berlin, 1994.

[13] E. D. Belokolos and I. M. Pershko, *The connection between the smoothness of the potential and the size of lacunae in the limit spectrum of the Schrödinger operator*, Ukrainian Math. J. **33** (1981), no. 5, 495–499. MR 0633738

[14] P. A. Binding and B. P. Rynne, *The spectrum of the periodic p-Laplacian*, J. Differential Equations **235** (2007), no. 1, 199–218. MR 2309572 (2008m:35093)

[15] G. D. Birkhoff, *Existence and oscillation theorems for a certain boundary value problem*, Trans. Amer. Math. Soc. **10** (1909), 259–270.

[16] L. E. Blumenson, *On the eigenvalues of Hill's equation*, Comm. Pure Appl. Math. **16** (1963), 261–266. MR 0157038 (28 #279)

[17] G. Borg, *Eine Umkehrung der Sturm-Liouvilleschen Eigenwertaufgabe. Bestimmung der Differentialgleichung durch die Eigenwerte*, Acta Math. **78** (1946), 1–96. MR 0015185 (7,382d)

[18] _____, *On a Liapounoff criterion of stability*, Amer. J. Math. **71** (1949), 67–70. MR 0028500 (10,456b)

[19] C. J. Bouwkamp, *A note on Mathieu functions*, Nederl. Akad. Wetensch., Proc. **51** (1948), 891–893=Indagationes Math. 10, 319–321 (1948). MR 0029008 (10,533b)

[20] H. Bremekamp, *Over de periodieke oplossingen der vergelijking van Mathieu*, Nieuw Archief voor Wiskunde **15**, 134–146.

[21] J. C. Bronski and Z. Rapti, *Counting defect modes in periodic eigenvalue problems*, SIAM J. Math. Anal. **43** (2011), no. 2, 803–827. MR 2784877 (2012g:34222)

[22] B. M. Brown and M. S. P. Eastham, *Titchmarsh's asymptotic formula for periodic eigenvalues and an extension to the p-Laplacian*, J. Math. Anal. Appl. **338** (2008), no. 2, 1255–1266. MR 2386494 (2008k:34333)

[23] _____, *An eigenvalue inequality for the periodic p-Laplacian and an explicit bound*, Comm. Appl. Nonlinear Anal. **16** (2009), no. 4, 61–71. MR 2591329 (2011a:34200)

[24] _____, *Generalised Meissner equations with an eigenvalue-inducing interface*, Operator theory: Advances and applications **219** (2012), 1–20.

[25] B. M. Brown, M. S. P. Eastham, A. M. Hinz, and K. M. Schmidt, *Distribution of eigenvalues in gaps of the essential spectrum of Sturm-Liouville operators—a numerical approach*, J. Comput. Anal. Appl. **6** (2004), no. 1, 85–95. MR 2223237 (2007a:65108)

[26] B. M. Brown and K. M. Schmidt, *On the HELP inequality for Hill operators on trees*, Spectral theory, function spaces and inequalities, Oper. Theory Adv. Appl., vol. 219, Birkhäuser/Springer Basel AG, Basel, 2012, pp. 21–36. MR 2848626

[27] M. Burnat, *The stability of eigenfunctions and the spectrum of Hill's equation*, Bull. Acad. Polon. Sci. Sér. Sci. Math. Astronom. Phys. **9** (1961), 795–797. MR 0132262 (24 #A2107)

[28] E. A. Coddington and N. Levinson, *Theory of ordinary differential equations*, McGraw-Hill Book Company, Inc., New York-Toronto-London, 1955. MR 0069338 (16,1022b)

[29] Y. Colin de Verdière and Th. Kappeler, *On double eigenvalues of Hill's operator*, J. Funct. Anal. **86** (1989), no. 1, 127–135. MR 1013936 (91b:34143)

[30] H. Coskun and B. J. Harris, *Estimates for the periodic and semi-periodic eigenvalues of Hill's equation*, Proc. Roy. Soc. Edinburgh Sect. A **130** (2000), no. 5, 991–998. MR 1800088 (2001i:34145)

[31] K. Daho and H. Langer, *Sturm-Liouville operators with an indefinite weight function: the periodic case*, Rad. Mat. **2** (1986), no. 2, 165–188. MR 873696 (88i:34048a)

[32] P. A. Deift and R. Hempel, *On the existence of eigenvalues of the Schrödinger operator $H - \lambda W$ in a gap of $\sigma(H)$*, Comm. Math. Phys. **103** (1986), no. 3, 461–490. MR 832922 (87k:35184)

[33] P. Djakov and B. Mityagin, *Smoothness of Schrödinger operator potential in the case of Gevrey type asymptotics of the gaps*, J. Funct. Anal. **195** (2002), no. 1, 89–128. MR 1934354 (2003k:34167)

[34] _____, *The asymptotics of spectral gaps of a 1D Dirac operator with cosine potential*, Lett. Math. Phys. **65** (2003), no. 2, 95–108. MR 2022123 (2006a:34237)

[35] _____, *Spectral gaps of the periodic Schrödinger operator when its potential is an entire function*, Adv. in Appl. Math. **31** (2003), no. 3, 562–596. MR 2006361 (2004k:47088)

[36] _____, *Multiplicities of the eigenvalues of periodic Dirac operators*, J. Differential Equations **210** (2005), no. 1, 178–216. MR 2114129 (2005k:34332)

[37] _____, *Simple and double eigenvalues of the Hill operator with a two-term potential*, J. Approx. Theory **135** (2005), no. 1, 70–104. MR 2151250 (2006a:34234)

[38] _____, *Asymptotics of instability zones of the Hill operator with a two term potential*, J. Funct. Anal. **242** (2007), no. 1, 157–194. MR 2274019 (2007j:34176)

[39] O. Došlý, *Rofe-Beketov's formula on time scales*, Comput. Math. Appl. **60** (2010), no. 8, 2382–2386. MR 2725329 (2012d:34233)

[40] B. A. Dubrovin, V. B. Matveev, and S. P. Novikov, *Nonlinear equations of Korteweg-de Vries type, finite zone linear operators and Abelian varieties*, Russian Math. Surveys **31** (1976), 59–146.

[41] M. S. P. Eastham, *Gaps in the essential spectrum associated with singular differential operators*, Quart. J. Math. Oxford Ser. (2) **18** (1967), 155–168. MR 0217361 (36 #451)

[42] _____, *On the limit points of the spectrum*, J. London Math. Soc. **43** (1968), 253–260. MR 0225027 (37 #624)

[43] _____, *On the gaps in the spectrum associated with Hill's equation*, Proc. Amer. Math. Soc. **21** (1969), 643–647. MR 0240377 (39 #1726)

[44] _____, *The Schrödinger equation with a periodic potential*, Proc. Roy. Soc. Edinburgh Sect. A **69** (1970/1971), 125–131. MR 0289966 (44 #7151)

[45] _____, *The first stability interval of the periodic Schrödinger equation*, J. London Math. Soc. (2) **4** (1972), 587–592. MR 0296533 (45 #5593)

[46] _____, *The periodic Schrödinger equation and the Brillouin zone*, J. London Math. Soc. (2) **5** (1972), 240–242. MR 0312065 (47 #627)

[47] _____, *Semi-bounded second-order differential operators.*, Proc. Roy. Soc. Edinburgh **72A** (1972/1973), 9–16.

[48] _____, *The spectral theory of periodic differential equations*, Scottish Academic Press, Edinburgh, 1973.

[49] _____, *Results and problems in the spectral theory of periodic differential equations*, Spectral theory and differential equations (Proc. Sympos., Dundee, 1974; dedicated to Konrad Jörgens), Springer, Berlin, 1975, pp. 126–135. Lecture Notes in Math., Vol. 448. MR 0404749 (53 #8549)

[50] _____, *Gaps in the essential spectrum of even-order self-adjoint differential operators*, Proc. London Math. Soc. (3) **34** (1977), no. 2, 213–230. MR 0437849 (55 #10770)

[51] M. S. P. Eastham, C. T. Fulton, and S. Pruess, *Using the SLEDGE package on Sturm-Liouville problems having nonempty essential spectra*, ACM Transactions on Mathematical Software **22** (1996), no. 4, 423–446.

[52] M. S. P. Eastham, Q. Kong, H. Wu, and A. Zettl, *Inequalities among eigenvalues of Sturm-Liouville problems*, J. Inequal. Appl. **3** (1999), no. 1, 25–43. MR 1731667 (2000k:34132)

[53] M. S. P. Eastham and K. M. Schmidt, *Absence of high-energy spectral concentration for Dirac systems with divergent potentials*, Proc. Roy. Soc. Edinburgh Sect. A **135** (2005), no. 4, 689–702. MR 2173335 (2006f:34163)

[54] R. F. Efendiev and H. D. Orudzhev, *Inverse wave spectral problem with discontinuous wave speed*, Zh. Mat. Fiz. Anal. Geom. **6** (2010), no. 3, 255–265, 337, 340. MR 2724827 (2011h:34020)

[55] I. J. Epstein, *Periodic solutions of systems of differential equations*, Proc. Amer. Math. Soc. **13** (1962), 690–694. MR 0144008 (26 #1556)

[56] A. Erdélyi, W. Magnus, F. Oberhettinger, and F. G. Tricomi, *Higher transcendental functions. Vol. III*, McGraw-Hill Book Company, Inc., New York-Toronto-London, 1955, Based, in part, on notes left by Harry Bateman. MR 0066496 (16,586c)

[57] V. A. Erovenko, *Instability zones of Hill's differential equation*, Differential Equations **23** (1975), 1413–1414.

[58] _____, *Lacunae in the spectrum of a fourth order differenetial operator with periodic coefficients*, Differencial′nye Uravnenija **11** (1975), no. 11, 1942–1948, 2106. MR 0397071 (53 #931)

[59] G. Floquet, *Sur les équations différentielle linéaires à coefficients périodiques*, Ann. Ecole Norm. ser **2** (1883), no. 12, 47–89.

[60] S. Gan and M. Zhang, *Resonance pockets of Hill's equations with two-step potentials*, SIAM J. Math. Anal. **32** (2000), no. 3, 651–664 (electronic). MR 1786162 (2001i:34085)

[61] F. Gesztesy, H. Holden, and W. Kirsch, *On energy gaps in a new type of analytically solvable model in quantum mechanics*, J. Math. Anal. Appl. **134** (1988), no. 1, 9–29. MR 958850 (90c:81032)

[62] F. Gesztesy and B. Simon, *On a theorem of Deift and Hempel*, Comm. Math. Phys. **116** (1988), no. 3, 503–505. MR 937772 (89g:35080)

[63] _____, *A short proof of Zheludev's theorem*, Trans. Amer. Math. Soc. **335** (1993), no. 1, 329–340. MR 1096260 (93c:34162)

[64] F. Gesztesy, B. Simon, and G. Teschl, *Zeros of the Wronskian and renormalized oscillation theory*, Amer. J. Math. **118** (1996), no. 3, 571–594. MR 1393260 (97g:34105)

[65] F. Gesztesy and M. Ünal, *Perturbative oscillation criteria and Hardy-type inequalities*, Math. Nachr. **189** (1998), 121–144. MR 1492926 (99a:34069)

[66] D. J. Gilbert and D. B. Pearson, *On subordinacy and analysis of the spectrum of one-dimensional Schrödinger operators*, J. Math. Anal. Appl. **128** (1987), no. 1, 30–56. MR 915965 (89a:34033)

[67] I. M. Glazman, *Direct methods of qualitative spectral analysis of singular differential operators*, Translated from the Russian by the IPST staff, Israel Program for Scientific Translations, Jerusalem, 1965, 1966. MR 0190800 (32 #8210)

[68] W. Goldberg, *On the determination of a Hill's equation from its spectrum*, J. Math. Anal. Appl. **51** (1975), no. 3, 705–723. MR 0372301 (51 #8517)

[69] _____, *Necessary and sufficient conditions for determining a Hill's equation from its spectrum*, J. Math. Anal. Appl. **55** (1976), no. 3, 549–554. MR 0427734 (55 #764)

[70] A. Grigis, *Estimations asymptotiques des intervalles d'instabilité pour l'équation de Hill*, Ann. Sci. École Norm. Sup. (4) **20** (1987), no. 4, 641–672. MR 932802 (89e:34056)

[71] G. Hamel, *Über die lineare Differentialgleichungen zweiter Ordnung mit periodischen Koeffizienten*, Math. Ann. **73** (1913), 371–412.

[72] G. H. Hardy, J. E. Littlewood, and G. Pólya, *Inequalities*, 2nd ed., Cambridge, 1959.

[73] E. M. Harrell, II, *On the effect of the boundary conditions on the eigenvalues of ordinary differential equations*, Contributions to analysis and geometry (Baltimore, Md., 1980), Johns Hopkins Univ. Press, Baltimore, Md., 1981, pp. 139–150. MR 648460 (83c:34031)

[74] B. J. Harris, *Gaps in the essential spectra of Schrödinger and Dirac operators*, J. London Math. Soc. (2) **18** (1978), no. 3, 489–501. MR 518234 (80a:35091)

[75] _____, *A systematic method of estimating gaps in the essential spectrum of self-adjoint operators*, J. London Math. Soc. (2) **18** (1978), no. 1, 115–132. MR 0492491 (58 #11607)

[76] _____, *On the essential spectrum of selfadjoint operators*, Proc. Roy. Soc. Edinburgh Sect. A **86** (1980), no. 3-4, 261–274. MR 592554 (81m:34023)

[77] _____, *On the spectra and stability of periodic differential equations*, Proc. London Math. Soc. (3) **41** (1980), no. 1, 161–192. MR 579720 (81g:34026)

[78] Ph. Hartman, *On the eigenvalues of differential equations*, Amer. J. Math. **73** (1951), 657–662. MR 0044703 (13,463c)

[79] _____, *Ordinary differential equations*, Classics in Applied Mathematics, vol. 38, Society for Industrial and Applied Mathematics (SIAM), Philadelphia, PA, 2002, Corrected reprint of the second (1982) edition [Birkhäuser, Boston, MA; MR0658490 (83e:34002)], With a foreword by Peter Bates. MR 1929104 (2003h:34001)

[80] O. Haupt, *Über lineare homogene Differentialgleichungen 2. Ordnung mit periodischen Koeffizienten*, Math. Ann. **79** (1918), 278–285.

[81] E. Heil, *Eigenvalue estimates for Hill's equation*, J. Differential Equations **18** (1975), 179–187. MR 0369784 (51 #6013)

Bibliography

[82] A. D. Hemery and A. P. Veselov, *Whittaker-Hill equation and semifinite-gap Schrödinger operators*, J. Math. Phys. **51** (2010), no. 7, 072108, 17. MR 2681073 (2011h:34190)

[83] R. Hempel, I. Herbst, A. M. Hinz, and H. Kalf, *Intervals of dense point spectrum for spherically symmetric Schrödinger operators of the type* $-\Delta +$ $\cos|x|$, J. London Math. Soc. (2) **43** (1991), no. 2, 295–304. MR 1111587 (92f:35109)

[84] E. Hille, *On the zeros of Mathieu functions*, Proc. London Math. Soc. **23** (1924), no. 2, 185–237.

[85] H. Hochstadt, *Asymptotic estimates for the Sturm-Liouville spectrum*, Comm. Pure Appl. Math. **14** (1961), 749–764. MR 0132863 (24 #A2699)

[86] _____, *Estimates on the stability intervals for Hill's equation*, Proc. Amer. Math. Soc. **14** (1963), 930–932. MR 0156023 (27 #5956)

[87] _____, *Function theoretic properties of the discriminant of Hill's equation*, Math. Z. **82** (1963), 237–242. MR 0156022 (27 #5955)

[88] _____, *On the asymptotic spectrum of Hill's equation*, Arch. Math. (Basel) **14** (1963), 34–38. MR 0147693 (26 #5207)

[89] _____, *A special Hill's equation with discontinuous coefficients*, Amer. Math. Monthly **70** (1963), 18–26. MR 0145145 (26 #2680)

[90] _____, *On the determination of a Hill's equation from its spectrum*, Arch. Rational Mech. Anal. **19** (1965), 353–362. MR 0181792 (31 #6019)

[91] _____, *On the determination of a Hill's equation from its spectrum. II*, Arch. Rational Mech. Anal. **23** (1966), 237–238. MR 0200520 (34 #411)

[92] _____, *The functions of mathematical physics*, Wiley-Interscience [A division of John Wiley & Sons, Inc.], New York-London-Sydney, 1971, Pure and Applied Mathematics, Vol. XXIII. MR 0499342 (58 #17241)

[93] _____, *On a Hill's equation with double eigenvalues*, Proc. Amer. Math. Soc. **65** (1977), no. 2, 373–374. MR 0445059 (56 #3404)

[94] _____, *On the width of the instability intervals of the Mathieu equation*, SIAM J. Math. Anal. **15** (1984), no. 1, 105–107. MR 728684 (85f:34087)

[95] E. L. Ince, *A proof of the impossibility of the coexistence of two Mathieu functions*, Proc. Camb. Phil. Soc. **21** (1922), 117–120.

[96] _____, *A linear differential equation with periodic coefficients.*, Proc. London Math. Soc. **23** (1923), 56–74.

[97] _____, *Ordinary differential equations*, Longmans, 1927.

[98] _____, *Periodic solutions of a linear differential equations of the second order with periodic coefficients*, Proc. Camb. Phil. Soc. **23** (1927), 44–46.

[99] _____, *Tables of the elliptic cylinder functions*, Proc. Roy. Soc. Edinburgh **52** (1931-2), 355–423.

[100] _____, *Further investigations into the periodic Lamé functions*, Proc. Roy. Soc. Edinburgh **60** (1940), 83–99. MR 0002400 (2,46d)

[101] V. A. Jakubovič, *A remark on the Floquet-Ljapunov theorem*, Vestnik Leningrad. Univ. **25** (1970), no. 1, 88–92. MR 0271443 (42 #6326)

[102] L. S. Jakupov, *Solutions in finite form for certain Hill's equations*, Soviet Math. Dokl. **9** (1968), 843–845. MR 0230961 (37 #6518)

[103] R. Johnson and J. Moser, *The rotation number for almost periodic potentials*, Comm. Math. Phys. **84** (1982), no. 3, 403–438. MR 667409 (83h:34018)

[104] K. Jörgens, *Die asymptotische Verteilung der Eigenwerte singulärer Sturm-Liouville-Probleme*, Ann. Acid. Sci. Fenn. Ser. A I No. **336/4** (1963), 24. MR 0190425 (32 #7838)

[105] I. Y. Karaca, *Asymptotic formulas for eigenvalues of a Hill's equation with piecewise constant coefficient*, Math. Proc. R. Ir. Acad. **109** (2009), no. 1, 19–34. MR 2475798 (2010j:34202)

[106] Y. E. Karpeshina, *Perturbation theory for the Schrödinger operator with a periodic potential*, Lecture Notes in Mathematics, vol. 1663, Springer-Verlag, Berlin, 1997. MR 1472485 (2000i:35002)

[107] T. Kato, *Note on the least eigenvalue of Hill's equation*, Quart. Appl. Math.. **10** (1952), 292–294.

[108] K. V. Khmelnytskaya and H. C. Rosu, *Spectral parameter power series representation for Hill's discriminant*, Ann. Physics **325** (2010), no. 11, 2512–2521. MR 2718558 (2011j:34045)

[109] S. V. Khryashchëv, *Discrete spectrum for a periodic Schroedinger operator perturbed by a decreasing potential*, Order, disorder and chaos in quantum systems (Dubna, 1989), Oper. Theory Adv. Appl., vol. 46, Birkhäuser, Basel, 1990, pp. 109–114. MR 1124656 (92f:34078)

[110] K. Klotter and G. Kotowski, *Über die Stabilität der Lösungen Hillscher Differentialgleichungen mit drei unabhängigen Parametern. Erste Mitteilung: Über die Gleichung* $y'' + (\lambda + \gamma_1 \cos x + \gamma_2 \cos 2x)y = 0$, Z. Angew. Math. Mech. **23** (1943), no. 3, 149–155.

[111] A. Kneser, *Untersuchungen über die reellen Nullstellen der Integrale linearer Differentialgleichungen*, Math. Ann. **42** (1893), no. 3, 409–435. MR 1510784

[112] Q. Kong, Q. Lin, H. Wu, and A. Zettl, *A new proof of the inequalities among Sturm-Liouville eigenvalues*, Panamer. Math. J. **10** (2000), no. 2, 1–11. MR 1754507 (2001c:34052)

[113] E. Korotyaev, *Estimates of periodic potentials in terms of gap lengths*, Comm. Math. Phys. **197** (1998), no. 3, 521–526. MR 1652779 (99h:34125)

[114] _____, *Inverse problem and estimates for periodic Zakharov-Shabat systems*, J. Reine Angew. Math. **583** (2005), 87–115. MR 2146853 (2005m:35313)

[115] _____, *Estimates for the Hill operator. II*, J. Differential Equations **223** (2006), no. 2, 229–260. MR 2214934 (2006k:34234)

[116] _____, *Conformal spectral theory for the monodromy matrix*, Trans. Amer. Math. Soc. **362** (2010), no. 7, 3435–3462. MR 2601596 (2011e:34188)

[117] S. Kotani, *Lyapunov exponents and spectra for one-dimensional random Schrödinger operators*, Random matrices and their applications (Brunswick, Maine, 1984), Contemp. Math., vol. 50, Amer. Math. Soc., Providence, RI, 1986, pp. 277–286. MR 841099 (88a:60116)

[118] H. A. Kramers, *Das Eigenwertproblem im eindimensionalen periodischen Kraftfeld*, Physica **2** (1935), 483–490.

[119] M. G. Kreĭn, *On the characteristic function $A(\lambda)$ of a linear canonical system of differential equations of second order with periodic coefficients*, Prikl. Mat. Meh. **21** (1957), 320–329. MR 0094502 (20 #1018)

[120] R. de L. Kronig and W. G. Penney, *Quantum mechanics in crystal lattices*, Proc. Roy. Soc. **130** (1931), 499–513.

[121] H. Krüger and G. Teschl, *Effective Prüfer angles and relative oscillation criteria*, J. Differential Equations **245** (2008), no. 12, 3823–3848. MR 2462706 (2009k:34077)

[122] M. R. S. Kulenović, *On a result of Etgen and Lewis*, Czechoslovak Math. J. **32(107)** (1982), no. 3, 373–376. MR 669778 (83k:34061)

[123] P. Kurasov and J. Larson, *Spectral asymptotics for Schrödinger operators with periodic point interactions*, J. Math. Anal. Appl. **266** (2002), no. 1, 127–148. MR 1876773 (2002j:34148)

[124] J. Lanke, *On the asymptotic distribution of eigenvalues of certain one-dimensional Sturm-Liouville operators*, Math. Scand. **9** (1961), 69–79. MR 0138017 (25 #1465)

[125] V. F. Lazutkin and T. F. Pankratova, *Asymptotics of the width of gaps in the spectrum of the Sturm-Liouville operator with a periodic potential*, Soviet Math. Dokl. **15** (1974), 649–653. MR 0355178 (50 #7654)

[126] B. M. Levitan and M. G. Gasymov, *Determination of a differential equation by two spectra*, Russian Math. Surveys **19** (1964), no. 2 (116), 3–63. MR 0162996 (29 #299)

[127] W. Magnus and S. Winkler, *Hill's equation*, Interscience Tracts in Pure and Applied Mathematics, No. 20, Interscience Publishers John Wiley & Sons New York-London-Sydney, 1966. MR 0197830 (33 #5991)

[128] V. A. Marchenko and I. V. Ostrowskii, *A characterization of the spectrum of Hill's operator*, Math. USSR Sb. **26** (1975), no. 4, 493–554.

[129] Z. Marković, *Sur la non-existence simultanée de deux fonctions de Mathieu*, Proc. Camb. Phil. Soc. **23** (1926), 203–205.

[130] _____, *Sur les solutions de l'équation différentielle linéaire du second ordre à coefficient périodique*, Proc. London Math. Soc **31** (1930), 417–438.

[131] E. Meissner, *Über Schüttelschwingungen in Systemen mit periodisch veränderlicher Elastizität*, Schweizer Bauzeitung **72** (1918), 95–98.

[132] J. Meixner and F. W. Schäfke, *Mathieusche Funktionen und Sphäroidfunktionen mit Anwendungen auf physikalische und technische Probleme*, Die Grundlehren der mathematischen Wissenschaften in Einzeldarstellungen mit besonderer Berücksichtigung der Anwendungsgebiete, Band LXXI, Springer-Verlag, Berlin, 1954. MR 0066500 (16,586g)

[133] H. Menken, *Accurate asymptotic formulas for eigenvalues and eigenfunctions of a boundary-value problem of fourth order*, Bound. Value Probl. (2010), Art ID 720235, 21. MR 2745088 (2011j:34284)

[134] V. A. Mikhailets and A. V. Sobolev, *Common eigenvalue problem and periodic Schrödinger operators*, J. Funct. Anal. **165** (1999), no. 1, 150–172. MR 1696696 (2000f:47041)

[135] W. E. Milne, *On the degree of convergence of expansions in an infinite interval*, Trans. Amer. Math. Soc. **31** (1929), no. 4, 907–918. MR 1501521

[136] R. A. Moore, *The least eigenvalue of Hill's equation*, J. Analyse Math. **5** (1956/57), 183–196. MR 0086200 (19,141f)

[137] J. Moser, *An example of a Schroedinger equation with almost periodic potential and nowhere dense spectrum*, Comment. Math. Helv. **56** (1981), no. 2, 198–224. MR 630951 (82k:34029)

[138] J. S. Muldowney, *Linear systems of differential equations with periodic solutions*, Proc. Amer. Math. Soc. **18** (1967), 22–27. MR 0206403 (34 #6222)

[139] E. Müller-Pfeiffer, *Eine Abschätzung für das Infimum des Spektrums des Hillschen Differentialoperators*, Publ. Math. Debrecen **25** (1978), no. 1-2, 35–40. MR 0493523 (58 #12519)

[140] _____, *Spectral theory of ordinary differential operators*, Ellis Horwood Ltd., Chichester, 1981, Translated from the German by the author, Translation edited by M. S. P. Eastham, Ellis Horwood Series in Mathematics and its Applications. MR 606197 (82f:34027)

[141] A. A. Ntinos, *Lengths of instability intervals of second order periodic differential equations*, Quart. J. Math. Oxford (2) **27** (1976), no. 107, 387–394. MR 0486767 (58 #6467)

[142] L. A. Pipes, *Matrix solution of equations of the Mathieu-Hill type*, J. Appl. Phys. **24** (1953), 902–910. MR 0056797 (15,128f)

[143] M. H. Protter and C. B. Morrey, *A first course in real analysis*, Springer-Verlag, New York, 1977, Undergraduate Texts in Mathematics. MR 0463372 (57 #3324)

[144] H. Prüfer, *Neue Herleitung der Sturm-Liouvilleschen Reihenentwicklung stetiger Funktionen*, Math. Ann. **95** (1926), 499–518.

[145] C. R. Putnam, *On the least eigenvalue of Hill's equation*, Quart. Appl. Math. **9** (1951), 310–314. MR 0044706 (13,463f)

[146] _____, *On the gaps in the spectrum of the Hill equation*, Quart. Appl. Math. **11** (1954), 496–498. MR 0059430 (15,528c)

[147] _____, *On the first stability interval of the Hill equation*, Quart. Appl. Math. **16** (1958), 421–422. MR 0100130 (20 #6564)

[148] _____, *On the stability intervals of the Hill equation*, J. Soc. Indust. Appl. Math. **7** (1959), 101–106. MR 0100131 (20 #6565)

[149] M. Reed and B. Simon, *Methods of modern mathematical physics. I*, second ed., Academic Press Inc. [Harcourt Brace Jovanovich Publishers], New York, 1980, Functional analysis. MR 751959 (85e:46002)

[150] F. S. Rofe-Beketov, *A perturbation of a Hill's operator, that has a first moment and a non-zero integral, introduces a single discrete level into each of the distant spectral lacunae*, Mathematical physics and functional analysis, No. 4 (Russian), Fiz.-Tehn. Inst. Nizkih Temperatur, Akad. Nauk Ukrain. SSR, Kharkov, 1973, pp. 158–159, 163. MR 0477257 (57 #16798)

[151] _____, *Spectral analysis of the Hill operator and of its perturbations*, Functional analysis, No. 9: Harmonic analysis on groups (Russian), Ul'janovsk. Gos. Ped. Inst., Ul'yanovsk, 1977, pp. 144–155. MR 0493524 (58 #12520)

[152] _____, *Generalization of the Prüfer transformation and the discrete spectrum in gaps of the continuous spectrum*, Spectral theory of operators (Proc. Second All-Union Summer Math. School, Baku, 1975) (Russian), "Èlm", Baku, 1979, pp. 146–153. MR 558545 (81i:34021)

[153] _____, *Kneser constants and effective masses for band potentials*, Soviet Phys. Dokl. **29** (1984), 391–393. MR 745043 (86c:34054)

[154] _____, *Spectrum perturbations, the Kneser-type constants and the effective masses of zones-type potentials*, Constructive Theory of Functions 84 , Sofia (1984), 757–766.

[155] _____, *On an estimate for the growth of solutions of canonical almost periodic systems*, Mat. Fiz. Anal. Geom. **1** (1994), no. 1, 139–148. MR 1483803 (98j:34058)

[156] F. S. Rofe-Beketov and A. M. Kholkin, *Spectral analysis of differential operators*, World Scientific Monograph Series in Mathematics, vol. 7, World Scientific Publishing Co. Pte. Ltd., Hackensack, NJ, 2005, Interplay between spectral and oscillatory properties, Translated from the Russian by Ognjen Milatovic and revised by the authors, With a foreword by Vladimir A. Marchenko. MR 2175241 (2006g:47070)

[157] N. S. Rosenfeld, *The eigenvalues of a class of singular differential operators*, Comm. Pure Appl. Math. **13** (1960), 395–405. MR 0118887 (22 #9656)

[158] K. M. Schmidt, *On the genericity of nonvanishing instability intervals in periodic Dirac systems*, Ann. Inst. H. Poincaré Phys. Théor. **59** (1993), no. 3, 315–326. MR 1276329 (95c:34145)

[159] _____, *Dense point spectrum and absolutely continuous spectrum in spherically symmetric Dirac operators*, Forum Math. **7** (1995), no. 4, 459–475. MR 1337149 (96f:47097)

[160] _____, *Oscillation of the perturbed Hill equation and the lower spectrum of radially periodic Schrödinger operators in the plane*, Proc. Amer. Math. Soc. **127** (1999), no. 8, 2367–2374. MR 1626474 (99j:34050)

[161] _____, *Critical coupling constants and eigenvalue asymptotics of perturbed periodic Sturm-Liouville operators*, Comm. Math. Phys. **211** (2000), no. 2, 465–485. MR 1754525 (2001i:34147)

[162] _____, *Relative oscillation–non-oscillation criteria for perturbed periodic Dirac systems*, J. Math. Anal. Appl. **246** (2000), no. 2, 591–607. MR 1761950 (2001d:34142)

[163] _____, *Eigenvalues in gaps of perturbed periodic Dirac operators: numerical evidence*, J. Comput. Appl. Math. **148** (2002), no. 1, 169–181, On the occasion of the 65th birthday of Professor Michael Eastham. MR 1946194 (2004d:34176)

[164] _____, *Eigenvalue asymptotics of perturbed periodic Dirac systems in the slow-decay limit*, Proc. Amer. Math. Soc. **131** (2003), no. 4, 1205–1214 (electronic). MR 1948112 (2003k:81056)

[165] J. Shi, *A new form of discriminant for Hill's equation*, Ann. Differential Equations **15** (1999), no. 2, 191–210. MR 1716215 (2000k:34047)

[166] J. Shi, M. Lin, and J. Chen, *The calculation for characteristic multiplier of Hill's equation*, Appl. Math. Comput. **159** (2004), no. 1, 57–77. MR 2094957

[167] B. Simon, *On the genericity of nonvanishing instability intervals in Hill's equation*, Ann. Inst. H. Poincaré Sect. A (N.S.) **24** (1976), no. 1, 91–93. MR 0473321 (57 #12992)

[168] S. G. Simonyan, *Asymptotic properties of the width of gaps in the spectrum of a Sturm-Liouville operator with a periodic analytic potential*, Differential Equations **6** (1970), 965–971.

[169] _____, *Some asymptotic estimates of a differential system with periodic perturbation*, Izv. Akad. Nauk Armjan SSR Ser. Mat. **11** (1976), 263–274, 289.

[170] A. V. Sobolev, *Weyl asymptotics for the discrete spectrum of the perturbed Hill operator*, Estimates and asymptotics for discrete spectra of integral and differential equations (Leningrad, 1989–90), Adv. Soviet Math., vol. 7, Amer. Math. Soc., Providence, RI, 1991, pp. 159–178. MR 1306512 (95i:34158)

[171] A. V. Sobolev and M. Solomyak, *Schrödinger operators on homogeneous metric trees: spectrum in gaps*, Rev. Math. Phys. **14** (2002), no. 5, 421–467. MR 1912093 (2004b:31009)

[172] R. Stadler and G. Teschl, *Relative oscillation theory for Dirac operators*, J. Math. Anal. Appl. **371** (2010), no. 2, 638–648. MR 2670140 (2011h:34192)

[173] V. M. Staržinskiĭ, *A survey of works on the conditions of stability of the trivial solution of a system of linear differential equations with periodic coefficients*, Amer. Math. Soc. Transl. (2) **1** (1955), 189–237. MR 0073774 (17,484c)

[174] J. J. Stoker, *Nonlinear vibrations in mechanical and electrical systems*, Interscience Publishers, Inc., New York, N.Y., 1950. MR 0034932 (11,666a)

[175] G. Stolz, *On the absolutely continuous spectrum of perturbed Sturm-Liouville operators*, J. Reine Angew. Math. **416** (1991), 1–23. MR 1099943 (92d:34161)

[176] M. J. O. Strutt, *Bounds for the characteristic values of Hill problems. I. Characteristic values with smallest moduli*, Nederl. Akad. Wetensch. Verslagen, Afd. Natuurkunde **52** (1943), 83–90. MR 0013488 (7,159g)

[177] _____, *Characteristic functions of Hill problems. II. Expansion formulas in series of periodic and of almost periodic characteristic functions*, Nederl. Akad. Wetensch. Verslagen, Afd. Natuurkunde **52** (1943), 584–591. MR 0013493 (7,161a)

[178] D. X. Sun, *A note for a second order periodic linear differential equation*, Commun. Nonlinear Sci. Numer. Simul. **15** (2010), no. 11, 3339–3348. MR 2646164 (2011g:34103)

[179] C. A. Swanson, *Comparison and oscillation theory of linear differential equations*, Academic Press, New York, 1968, Mathematics in Science and Engineering, Vol. 48. MR 0463570 (57 #3515)

[180] G. Teschl, *Renormalized oscillation theory for Dirac operators*, Proc. Amer. Math. Soc. **126** (1998), no. 6, 1685–1695. MR 1443411 (98g:34137)

[181] E. C. Titchmarsh, *Theory of functions*, Clarendon Press, Oxford, 1939.

[182] _____, *Eigenfunction problems with periodic potentials*, Proc. Roy. Soc. London. Ser. A. **203** (1950), 501–514. MR 0039151 (12,502d)

[183] _____, *Eigenfunction expansions associated with second-order differential equations. part 2*, Clarendon Press, Oxford, 1958. MR 0176151 (31 #426)

[184] _____, *Some eigenfunction expansion formulae*, Proc. London Math. Soc. (3) **11** (1961), 159–168. MR 0123053 (23 #A384)

[185] _____, *Eigenfunction expansions associated with second-order differential equations, part 1*, Second Edition, Clarendon Press, Oxford, 1962. MR 0019765 (8,458d)

[186] V. A. Tkačenko, *Spectral analysis of the one-dimensional Schrödinger operator with a periodic complex-valued potential*, Soviet Math. Doklady **5** (1964), 413–415. MR 0163165 (29 #468)

[187] _____, *Spectrum of 1-d selfadjoint periodic differential operator of order 4*, Advances in differential equations and mathematical physics (Birmingham, AL, 2002), Contemp. Math., vol. 327, Amer. Math. Soc., Providence, RI, 2003, pp. 331–340. MR 1991552 (2004f:47066)

[188] E. Trubowitz, *The inverse problem for periodic potentials*, Comm. Pure Appl. Math. **30** (1977), no. 3, 321–337. MR 0430403 (55 #3408)

[189] P. Ungar, *Stable Hill equations*, Comm. Pure Appl. Math. **14** (1961), 707–710. MR 0176148 (31 #423)

[190] E. R. van Kampen and A. Wintner, *On an absolute constant in the theory of variational stability*, Amer. J. Math. **59** (1937), 270–274.

[191] S. Wallach, *The spectra of periodic potentials*, Amer. J. Math. **70** (1948), 842–848. MR 0027924 (10,376b)

[192] W. Walter, *Differential and integral inequalities*, Translated from the German by Lisa Rosenblatt and Lawrence Shampine. Ergebnisse der Mathematik und ihrer Grenzgebiete, Band 55, Springer-Verlag, New York, 1970. MR 0271508 (42 #6391)

[193] J. Weidmann, *Oszillationsmethoden für Systeme gewöhnlicher Differentialgleichungen*, Math. Z. **119** (1971), 349–373. MR 0285758 (44 #2975)

[194] _____, *Spectral theory of ordinary differential operators*, Lecture Notes in Mathematics, vol. 1258, Springer-Verlag, Berlin, 1987. MR 923320 (89b:47070)

[195] _____, *Uniform nonsubordinacy and the absolutely continuous spectrum*, Analysis **16** (1996), no. 1, 89–99. MR 1384355 (96m:34154)

[196] _____, *Spectral theory of Sturm-Liouville operators approximation by regular problems*, Sturm-Liouville theory, Birkhäuser, Basel, 2005, pp. 75–98. MR 2145078

[197] R. Weikard, *On Hill's equation with a singular complex-valued potential*, Proc. London Math. Soc. (3) **76** (1998), no. 3, 603–633. MR 1620492 (99e:34129)

[198] H. Weyl, *Über gewöhnliche Differentialgleichungen mit Singularitäten und die zugehörigen Entwicklungen willkürlicher Funktionen*, Math. Ann **68** (1910), 220–2695.

[199] A. Wintner, *The adiabatic linear oscillator*, Amer. J. Math. **68** (1946), 385–397. MR 0016811 (8,71b)

[200] _____, *Stability and spectrum in the wave mechanics of lattices*, Physical Rev. (2) **72** (1947), 81–82. MR 0020939 (8,615f)

[201] _____, *On the location of continuous spectra*, Amer. J. Math. **70** (1948), 22–30. MR 0023995 (9,435k)

[202] _____, *On the non-existence of conjugate points*, Amer. J. Math. **73** (1951), 368–380. MR 0042005 (13,37d)

[203] V. A. Yakubovich and V. M. Starzhinskii, *Linear differential equations with periodic coefficients. 1, 2*, Halsted Press [John Wiley & Sons] New York-Toronto, Ont., 1975, Translated from Russian by D. Louvish. MR 0364740 (51 #994)

[204] I. Yaslan and G. Sh. Guseinov, *On Hill's equation with piecewise constant coefficient*, Turkish J. Math. **21** (1997), no. 4, 461–474. MR 1621375 (99c:34201)

[205] K. Yoshitomi, *Coexistence problems for Hill's equations with three-step potentials*, Publ. Math. Debrecen **66** (2005), no. 3-4, 427–437. MR 2137779 (2005m:34198)

[206] _____, *Spectral gaps of the one-dimensional Schrödinger operators with periodic point interactions*, Hokkaido Math. J. **35** (2006), no. 2, 365–378. MR 2254656 (2007d:34160)

[207] _____, *Spectral gaps of the Schrödinger operators with periodic δ'-interactions and Diophantine approximations*, Math. Proc. Cambridge Philos. Soc. **143** (2007), no. 1, 185–199. MR 2340983 (2008g:34228)

[208] L. B. Zelenko, *Asymptotic distribution of the eigenvalues in a lacuna of the continuous spectrum of a perturbed Hill operator*, Mat. Zametki **20** (1976), no. 3, 341–350. MR 0430404 (55 #3409)

[209] M. Zhang, *The rotation number approach to eigenvalues of the one-dimensional p-Laplacian with periodic potentials*, J. London Math. Soc. (2) **64** (2001), no. 1, 125–143. MR 1840775 (2002e:35188)

[210] M. Zhang and S. Gan, *Constructing resonance calabashes of Hill's equations using step potentials*, Math. Proc. Cambridge Philos. Soc. **129** (2000), no. 1, 153–164. MR 1757785 (2001c:34181)

Index

ω-twisted boundary-value problem, 21, 114
    bound for least eigenvalue, 61
    eigenvalue comparison, 59
p-Laplacian, 64, 108, 109

absolutely continuous function, 2
absolutely continuous spectrum, 136, 148, 155
    in stability intervals, 140, 166
adiabatic limit, 170
angular momentum, 172
Arzelà-Ascoli theorem, 157
asymptotic expansion of solutions, 91, 93, 99

Baire category, first, 105
Baire-almost all, 105
boundary-value problem, 21
    on multi-period interval, 23
    on shifted period interval, 45, 94
    with separated boundary conditions, 37, 114
        eigenvalue comparison, 57
        eigenvalues, 37
    with equal boundary conditions, 45
bounded solutions, 162

Chaplygin inequality, 35
coexistence, 10, 22
    absence of, 19
comparison
    of eigenvalues, 57, 59
    Sturm, 36

continuous spectrum, 136, 152
critical
    constant, 181, 183, 194
    decay rate, 175

density of states, integrated, 44
Dirac operator
    full-line, 148
    half-line, 132
Dirac system, 13
Dirichlet
    boundary condition, 37
    eigenvalue, 37
    integral, 58
disconjugate equation, 47
discriminant, 9, 14
    derivative, 16, 18, 180
    example, 52
    qualitative behaviour, 17
    second derivative, 17
    under perturbation, 106

eigenfunction, 21, 37
    of half-line operator, 133
    orthogonality, 22
eigenvalue
    countability, 135
    in instability interval, 136
        asymptotics, 173, 192, 195
    of boundary-value problem, 21, 37
    of half-line operator, 133
    of periodic boundary-value problem, 22, 41

of semi-periodic boundary-value problem, 22, 41
  relation to spectral function, 133, 147
  stability under perturbation, 119
eigenvalues
  in instability interval, 168, 169
    asymptotics, 171
enumeration of closed instability intervals, 40
equation
  almost periodic, 64, 111
  Hill, 13
    step-function example, 52, 56
    with even coefficients, 55
  inhomogeneous, 114
  limit-periodic, 111
  Mathieu, 19
    extension of, 28, 56, 110
    length of instability intervals, 86
  Meissner, 11, 65
  Prüfer, 32
  Prüfer-type, 33
  Riccati, 33
  Schrödinger, 13
  Sturm-Liouville, 13
essential spectrum, 152

first category set, 105
Floquet
  exponent, 6
    asymptotics, 184
    example for, 8
  multiplier, 6, 9, 138
    example for, 12
  solution, 6, 18, 44
    asymptotics, 184
formula
  d'Alembert, 24
  Liouville, 3, 9
  Rofe-Beketov, 4, 10, 25
Fourier method, 20, 48, 73, 87

full-line operator, 147
  self-adjointness, 147
  spectrum, 148
fundamental matrix, 2
  canonical, 3, 5

general hypotheses, 13
generalised Fourier transform, 58, 120, 128, 130, 142, 146
  expansion formula, 118
  is isometry, 128
Glazman decomposition, 193
Green's function, 115
Green's identity, 121

half-line operator, 131
  self-adjointness, 133
  spectrum, 133
harmonic–arithmetic mean inequality, 50
Helly's theorems, 158
Hilbert space, 114
Hilbert-Schmidt theorem, 117
Hill discriminant, 9, 14
  derivative, 16, 18, 180
  example, 52
  qualitative behaviour, 17
  second derivative, 17
  under perturbation, 106
homogenisation, 170

inner product, 114
instability, 11
instability interval, 15
  all non-vanishing, 105
  all odd vanishing, 102
  all vanishing, 98
  almost all vanishing, 95, 101
  and (semi-)periodic eigenvalues, 41
  asymptotic length, 79
  bound on length, 84
  closed, 18, 40
    numbering, 40

Index

consecutive vanishing, 107
eigenvalue in, 136
eigenvalues in closed, 40, 45
exponential bound on length, 80
is spectral gap, 136
length, 151
length sequence, 110
of Mathieu equation, 86
vanishing, 21, 94
integrated density of states, 44
interlacing of eigenvalues, 41

Kepler transformation, 34, 177, 187
Kotani's theorem, 156

Lebesgue decomposition, 135, 158
limit-circle case, 123, 158
limit-point case, 123, 155
Liouville's formula, 3, 9
Liouville-Green transformation, 68

matrix
    diagonal partitioned, 6
    inverse of 2×2, 4, 176
    Jordan normal form, 6
    operator norm, 162
matrix exponential, 6
meagre set, 105
monodromy matrix, 5
    eigenvectors, 138
    example, 11
monotonic
    dependence of Prüfer angle on λ, 36
    discriminant, 15
multiplicity, total, 151

Neumann
    boundary condition, 37
    eigenvalue, 37
normalised eigenfunction, 21
nowhere dense set, 105

one-way rule, 38
operator core, 131

orthogonality of eigenfunctions, 22
oscillatory equation, 47, 195

Parseval identity, 59, 118, 120, 128, 146
Pauli matrices, 14
Peano inequality, 35
periodic
    all solutions, 8, 10
    boundary-value problem, 21
    eigenvalue, 22, 41, 53
        asymptotics, 70, 73, 76
        bound for least, 62
        upper bound, 47
    function, 5, 6, 10
    solution, 11
point interaction, 109
point spectrum, 136
    dense, 193
Prüfer angle asymptotics, 40, 42
Prüfer transformation, 32, 68, 176
pure point spectrum, 136

quantum mechanics, 13, 14, 43
quasi-derivative, 13
quasi-momentum, 43, 64

radially periodic Schrödinger operator, 193, 194
relative oscillation theorem, 39, 153, 159
resolvent operator, 115
    for ω-twisted boundary-value problem, 117
    for separated boundary conditions, 115
    norm of, 118
Riemann-Lebesgue lemma, 69, 93
rotation number, 42, 64, 171, 172
    derivative, 44

semi-periodic
    boundary-value problem, 21
    eigenvalue, 22, 41, 54

asymptotics, 70, 74
  upper bound, 48
function, 5
solution, 11
semiclassical asymptotics, 195
separated boundary conditions, 37
singular continuous spectrum, 136
singular sequence, 135, 148
slow variation limit, 170
spectral averaging, 155, 158
spectral function, 120, 127
  absolutely continuous in stability interval, 140
  matrix-valued, 142
  relation to spectrum, 133, 147
spectral subspace, 151
spectrum, 133
  relation to spectral function, 133, 147
stability, 11
  conditional, 11, 22
stability interval, 15
  and (semi-)periodic eigenvalues, 41
  and multi-period boundary-value problem, 24
  as set of twisted eigenvalues, 23
  bound for length, 61
  is absolutely continuous spectrum, 140, 148, 166
  lower bound for upper end, 49
  uniform boundedness of solutions, 137
Stieltjes inversion formula, 127, 145
Sturm comparison, 36
Sturm-Liouville operator
  full-line, 148
  half-line, 131
Sturm-Liouville system, 13
subordinacy, 159

theorem
  existence and uniqueness, 2
  relative oscillation, 39

Sturm comparison, 36
Titchmarsh-Weyl m-function, 124, 143, 158
total spectral multiplicity, 151
tree, regular, 158

uniform non-subordinacy, 155

variation of constants, 3, 91, 114, 176

Weierstrass elliptic function, 102, 111
Weyl circle, 122
Weyl's alternative, 124
Wirtinger inequality, 50, 65
Wronskian, 3

zeros of eigenfunctions, 37, 46

 **Birkhäuser** | **www.birkhauser-science.com**

Operator Theory: Advances and Applications (OT)

This series is devoted to the publication of current research in operator theory, with particular emphasis on applications to classical analysis and the theory of integral equations, as well as to numerical analysis, mathematical physics and mathematical methods in electrical engineering.

Edited by
Joseph A. Ball (Blacksburg, VA, USA), Harry Dym (Rehovot, Israel),
Marinus A. Kaashoek (Amsterdam, The Netherlands), Heinz Langer (Vienna, Austria),
Christiane Tretter (Bern, Switzerland)

■ **OT 229: Almeida, A. / Castro, L. / Speck, F.-O.** (Eds.), Advances in Harmonic Analysis and Operator Theory. The Stefan Samko Anniversary Volume (2013).
ISBN 978-3-0348-0515-5

■ **OT 228: Karlovich, Y.I. / Rodino, L. / Silbermann, B. / Spitkovsky, I.M.** (Eds.), Operator Theory, Pseudo-Differential Equations, and Mathematical Physics (2013).
ISBN 978-3-0348-0536-0

■ **OT 227: Janas, J. / Kurasov, P. / Laptev, A. / Naboko, S.** (Eds.), Operator Methods in Mathematical Physics. Conference on Operator Theory, Analysis and Mathematical Physics (OTAMP) 2010, Bedlewo, Poland (2013).
ISBN 978-3-0348-0530-8

■ **OT 226: Alpay, D. / Kirstein, B.**, Interpolation, Schur Functions and Moment Problems II (2012).
ISBN 978-3-0348-0427-1

■ **OT 225: Sakhnovich, L. A.**, Levy Processes, Integral Equations, Statistical Physics: Connections and Interactions (2012).
ISBN 978-3-0348-0355-7

■ **OT 224: Benguria, R. / Friedman, E. / Mantoiu, M.** (Eds.), Spectral Analysis of Quantum Hamiltonians. Spectral Days 2010 (2012).
ISBN 978-3-0348-0413-4

■ **OT 223: Jacob, B. / Zwart, H. J.**, Linear Port-Hamiltonian Systems on Infinite-dimensional Spaces (2012).
ISBN 978-3-0348-0398-4

■ **OT 222: Dym, H. / de Oliveira, M.C. / Putinar, M.** (Eds.), Mathematical Methods in Systems, Optimization, and Control. Festschrift in Honor of J. William Helton (2012).
ISBN 978-3-0348-0262-8

■ **OT 221: Arendt, W. / Ball, J.A. / Behrndt, J. / Förster, K.-H. / Mehrmann, V. / Trunk, C.** (Eds.), Spectral Theory, Mathematical System Theory, Evolution Equations, Differential and Difference Equations. IWOTA 10 (2012).
ISBN 978-3-0348-0262-8

■ **OT 220: Ball, J.A. / Curto, R.E. / Grudsky, S.M. / Helton, J.W.; Quiroga-Barranco, R. / Vasilevski, N.L.** (Eds.), Recent Progress in Operator Theory and Its Applications (2012).
ISBN 978-3-0348-0345-8

■ **OT 219: Brown, B.M. / Lang, J. / Wood, I.G.** (Eds.), Spectral Theory, Function Spaces and Inequalities. New Techniques and Recent Trends (2012).
ISBN 978-3-0348-0262-8

■ **OT 218: Dym, H. / Kaashoek, M.A. / Lancaster, P. / Langer, H. / Lerer, L.** (Eds.) A Panorama of Modern Operator Theory and Related Topics. The Israel Gohberg Memorial Volume (2012).
ISBN 978-3-0348-0220-8

■ **OT 217: Arlinskii, Y. / Belyi, S. / Tsekanovskii, E.**, Conservative Realizations of Herglotz–Nevanlinna Functions (2011).
ISBN 978-3-7643-9995-5

MIX
Papier aus verantwortungsvollen Quellen
Paper from responsible sources
FSC® C105338

If you have any concerns about our products,
you can contact us on
ProductSafety@springernature.com

In case Publisher is established outside the EU,
the EU authorized representative is:
**Springer Nature Customer Service Center GmbH
Europaplatz 3, 69115 Heidelberg, Germany**

Printed by Libri Plureos GmbH
in Hamburg, Germany